秋霞圃池上草堂
Chishang Cottage, Qiuxia Garden

图书在版编目（CIP）数据

文物视角中的江南. 上海古园林：汉文、英文 / 上海市文物保护研究中心编著. -- 上海：同济大学出版社，2025.4. -- (建筑可阅读书系). -- ISBN 978-7-5765-1587-9

Ⅰ. TU-092.2

中国国家版本馆 CIP 数据核字第 20256W9Q39 号

文物视角中的江南
上海古园林

上海市文物保护研究中心　编著

出 品 人	金英伟
责任编辑	由爱华　金　言
责任校对	徐逢乔
书籍设计	张　微
出版发行	同济大学出版社 www.tongjipress.com.cn
	（地址：上海市四平路 1239 号　邮编：200092　电话：021-65985622）
经　　销	全国新华书店
印　　刷	上海雅昌艺术印刷有限公司
开　　本	787mm×1092mm　1/36
印　　张	7
字　　数	175 000
版　　次	2025 年 4 月第 1 版
印　　次	2025 年 4 月第 1 次印刷
书　　号	ISBN 978-7-5765-1587-9
定　　价	68.00 元

本书若有印装问题，请向本社发行部调换
版权所有　侵权必究

豫园鱼乐榭
Yule Gazebo, Yu Garden

古猗园鸳鸯湖
Yuanyang Lake, Guyi Garden

this project to life.

We hope this book inspires readers to explore Shanghai's classical gardens, appreciate the beauty of Jiangnan's cultural heritage, and contribute to the preservation of these cultural treasures.

Despite our best efforts, omissions or inadequacies may inevitably remain in this book. We sincerely welcome critiques and corrections from our readers.

Shanghai Cultural Heritage Conservation and Research Center
November 11, 2024

Epilogue

The Jiangnan region is renowned for its rich and enduring cultural heritage, a cornerstone of Chinese traditional culture and a shared identity for the Yangtze River Delta. This heritage forms a key foundation for Shanghai's vibrant urban culture.

Since 2018, the Shanghai Municipal Administration of Culture and Tourism (Shanghai Municipal Administration of Cultural Heritage) launched the "Stories of Shanghai Architecture" initiative, which has been warmly received by the public. Building on this success, the Shanghai Cultural Heritage Conservation and Research Center created the *Jiangnan from the Perspective of Cultural Heritage* volumes of Stories of Shanghai Architecture series to celebrate Shanghai's heritage and enhance its cultural brand. This book, the second in the series, delves into the richness of Jiangnan culture through the lens of immovable cultural heritage (the first book is *Jiangnan from the Perspective of Cultural Relics: Ancient Towns in Shanghai*, 2024, which interpreted 11 national famous historical and cultural towns in Shanghai).

Shanghai's ancient gardens, which can be traced back to before the Tang dynasty, and Northern and Southern dynasties (220–589) and flourished during the Ming and Qing dynasties (1368–1911), have continued to evolve into the modern era. These gardens exemplify the transformation of Jiangnan classical gardens, transitioning from private retreats for scholars to modern public spaces. Reflecting the distinctive characteristics of their times, they showcase Shanghai's deep cultural roots while also emphasizing the city's inclusivity, diversity, and innovative spirit.

As this book comes to completion, we extend our heartfelt gratitude to the leaders who guided us, the experts and scholars who offered invaluable insights, and the staff of the gardens, cultural heritage departments, for their dedicated efforts. We gratefully acknowledge the meticulously surveyed drawings provided by the faculty and students of the Department of Architecture, College of Architecture and Urban Planning, Tongji University, as well as the calligraphy and painting images from the Shanghai History Museum (Shanghai Revolution Museum). We also thank Tongji University Press for their meticulous work, along with the editors and designers whose creativity brought

后记

江南文化底蕴悠长、内涵丰富，既是中华优秀传统文化的重要组成部分，也是长三角区域共同的文化标识，更是上海城市文化的重要根源。

自2018年起，上海市文化和旅游局（上海市文物局）着力推进"建筑可阅读"工作，深受欢迎。为推进"建筑可阅读"深度发展，打响上海文化品牌，弘扬城市文脉，上海市文物保护研究中心推出了建筑可阅读书系之文物视角中的江南系列，尝试从不可移动文物的角度阐释江南文化的丰富内涵，于2024年出版《文物视角中的江南：上海古镇》，解读上海11处中国历史文化名镇，本书是该系列丛书的第二本。

上海古园林萌芽于唐代之前，兴盛于明清时期，在近代依旧有所发展，是江南古园林流变的范式之一。从文人园林到近代园林，再到城市公共空间，上海古园林走出了一条具有时代与地域特点的道路，既展现了上海的文化底蕴，是江南文化重要的物质载体，又反映出了这座城市海纳百川、兼容并包、求新立异的特征。

成书之际，感谢各级领导的关怀，感谢各专家学者的帮助，感谢各园林方以及各区文物管理部门的大力支持，感谢同济大学建筑与城市规划学院建筑系师生提供的测绘图纸，感谢上海市历史博物馆（上海革命历史博物馆）提供的书画图片，感谢同济大学出版社的严谨刊布以及本书的责任编辑、美术编辑的辛勤付出。期望本书的出版，能带动大家探寻江南园林，感知园林艺术，体味江南文化，关注文化遗产保护与传承。

本书难免存在疏漏或不足之处，敬请读者批评指正。

编者

2024年11月11日

课植园后花厅出口
Houhua Hall Exit, Kezhi Garden

参考文献
Reference

[1]《上海园林志》编纂委员会.上海园林志[M].上海:上海社会科学院出版社,2000.

[2] 常青,朱宇晖,同济大学建筑与城规学院.中国古建筑测绘大系·祠庙建筑与园林建筑:上海庙园[M].北京:中国建筑工业出版社,2022.

[3] 陈从周.说园[M].上海:同济大学出版社,2007.

[4] 陈菊兴.青浦碑刻[Z].上海:青浦博物馆内部资料,1998.

[5] 陈业伟.豫园[M].上海:上海文化出版社,2009.

[6] 段建强.陈从周先生与豫园修复研究[C]// 中国建筑学会建筑史学分会,华南理工大学建筑学院.《营造》第五辑——第五届中国建筑史学国际研讨会会议论文集(下).2010.

[7] 阮仪三.朱家角[M].杭州:浙江摄影出版社,2004.

[8] 上海古猗园.古猗园志[M].上海:上海文化出版社,2018.

[9] 上海市地方志办公室,上海市闵行区地方志办公室.上海县卷[M].上海:上海古籍出版社,2015.

[10] 上海市青浦区博物馆.青浦望族[M].上海:上海人民出版社,2016.

[11] 同济大学建筑与城市规划学院景观学系.陈从周造园三章[M].上海:同济大学出版社,2018.

[12] 童寯.江南园林志[M].2版.北京:中国建筑工业出版社,2014.

[13] 张承先.南翔镇志[M].程攸熙,订.朱瑞熙,标点.上海:上海古籍出版社,2003.

[14] 朱家角镇地方志编纂委员会.朱家角乡志[M].香港:新大陆出版社有限公司,2007.

[15] 朱宇晖,路秉杰.上海传统园林研究[M].上海:同济大学出版社,2019.

[16] 朱宇晖.江南名园指南(上)[M].上海:上海科学技术出版社,2002.

古猗园荷风竹露亭
Hefengzhulu Pavilion, Guyi Garden

图片来源
Image Source

豫园测绘图
常青，朱宇晖，同济大学建筑与城规学院. 中国古建筑测绘大系·祠庙建筑与园林建筑：上海庙园 [M]. 北京：中国建筑工业出版社，2022：25.

秋霞圃测绘图
常青，朱宇晖，同济大学建筑与城规学院. 中国古建筑测绘大系·祠庙建筑与园林建筑：上海庙园 [M]. 北京：中国建筑工业出版社，2022：161.

古猗园测绘图
同济大学建筑与城规学院提供。同济大学18级4班2021年7月测绘，领队老师：王红军、刘涤宇、李渼、汤众；测量制图：汪志勇、黎嘉颖、张明慧。

醉白池内园测绘图
常青，朱宇晖，同济大学建筑与城规学院. 中国古建筑测绘大系·祠庙建筑与园林建筑：上海庙园 [M]. 北京：中国建筑工业出版社，2022：321.

曲水园测绘图
常青，朱宇晖，同济大学建筑与城规学院. 中国古建筑测绘大系·祠庙建筑与园林建筑：上海庙园 [M]. 北京：中国建筑工业出版社，2022：203.

课植园图
朱家角课植园提供，宋婧璇重绘

清同治《上海县志》城隍庙图
上海市地方志办公室，上海市闵行区地方志办公室. 上海县卷 [M]. 上海：上海古籍出版社，2015.

清末的湖心亭一带
常青，朱宇晖，同济大学建筑与城规学院. 中国古建筑测绘大系·祠庙建筑与园林建筑：上海庙园 [M]. 北京：中国建筑工业出版社，2022：18.

吴友如《豫园宴乐图》；任伯年为豫园点春堂所作的《观剑图》
上海市历史博物馆（上海革命历史博物馆）提供

明龚弘像（清代程祖庆《练川名人画像》）；20世纪60年代初的秋霞圃（陈从周摄）；清嘉庆《嘉定县志》邑庙灵苑图；秋霞圃平面图，20世纪60年代初（陈从周绘）；20世纪50年代的屏山堂（上海翥云艺术博物馆藏）；20世纪70年代的屏山堂（上海翥云艺术博物馆藏）；1621年，明代书法家娄坚题"涉趣桥"（1838年复刻）；《练师吕绍宾六十寿庆记》碑文，1923
上海市嘉定博物馆提供

古猗园鸟瞰图；20世纪初的不系舟；1911年的不系舟；20世纪30年代的不系舟；20世纪30年代小松冈及周边景象；1959年重建后的不系舟；小松冈
上海古猗园提供

《嘉定县志》古猗园图，清嘉庆十六年（1811）
北京大学图书馆. 北京大学图书馆藏稀见方志丛刊 [M]. 北京：国家图书馆出版社，2013.

《南翔镇志》南翔镇图，清嘉庆十一年（1806）
张承先. 南翔镇志 [M]. 程攸熙，订. 朱瑞熙，标点. 上海：上海古籍出版社，2003.

《江南园林志》古猗园平面图（20世纪30年代童寯先生绘）
童寯. 江南园林志 [M]. 2版. 北京：中国建筑工业出版社，1984.

醉白池鸟瞰图；清朝末年的醉白池；池上草堂（摄于20世纪50年代）；原大门旧貌（摄于20世纪80年代）；池上草堂侧房；百年牡丹；莲叶东南榭（大湖亭）、池南廊壁上的《云间邦彦画像》；疑舫；"育婴堂奉文保管界"碑
上海醉白池公园提供

曲水园荷花池旧照；曲水园图（《青浦县志》灵园图）；曲水园示意图；春天的曲水园
上海曲水园提供

上海古园林分布示意图
苏诗淼绘

其他未标注图片均由许一凡拍摄

秋霞圃园景
The scenery of Qiuxia Garden

倒挂狮子亭
Daoshi Pavilion

课植精神
The Spirit of Studying and Tilling

古园林具有悠久的历史，是我国重要的文化遗产之一，在世界园林中独树一帜，也是中国传统文化的重要组成部分，凝聚着中华民族的智慧与审美。课植园有别于沪上其他园林，其独特之处在于"植园"。"植园"中的雅事、农事活动充分展现了中国传统社会中的勤耕苦读、田野情趣。如今，随着上海文旅事业的繁荣，课植园这座百年沪上名园的"课植"精神必会绵延流传，熠熠生辉。

别有洞天
A hidden and enchanting place

参观指南 Visiting Guide
开放时间：08:30—16:30。
交通信息：地铁 17 号线至朱家角站。

Opening hours: 08:30-16:30.
Traffic information: Metro Line 17 to Zhujiajiao Station.

Classical gardens celebrated for their rich history, are a significant part of China's cultural heritage. Their distinctive style that distinguishes them in the world of landscape design. These gardens are an integral element of traditional Chinese culture, embodying the wisdom and aesthetic ideals of the Chinese people. Kezhi Garden, in particular, stands out among Shanghai's gardens due to its unique focus on the Zhi (tilling) section. The garden's agricultural and scholarly activities vividly illustrate traditional values of diligent study, hard work, and the simple beauty of rural life in ancient China. Today, as Shanghai's cultural and tourism industries prosper, the enduring spirit of "Kezhi" continues to thrive in this century-old garden, affirming its lasting cultural significance.

---

1 朱家角镇地方志编纂委员会. 朱家角乡志 [M]. 香港：新大陆出版社有限公司，2007.
2 上海市青浦区博物馆. 青浦望族 [M]. 上海：上海人民出版社，2016.

山洞清泉
Mountain stream with clear springs

as county magistrate, visited Jingting Town to inspect an insect infestation and stayed at Ma Weiqi's Kezhi Garden. Over time, he came to admire Ma's values and composed the biography, which Ma later commissioned Zhou Meigu to engrave on a bluestone stele for public display. Originally placed beside the garden's rockery, the stele measured 78 centimeters high and 163 centimeters wide. It was later broken into three pieces, but its message endures.

In 2012, during Kezhi Garden's centennial celebrations, the stele fragments were rediscovered and studied. Although much of the inscription has faded and become difficult to read, the full text survives in the *Ma Family Genealogy*. It opens with a quote from the *Analects*, specifically "Fan Chi Asks about Farming," emphasizing that farming requires practical, hands-on experience. The inscription describes Ma Weiqi, the garden's owner, who was a native of Jingting Township in Kunshan. Even in his seventies, Ma, with his white hair, maintained a youthful appearance, looking more like a man in his forties or fifties, and moved with notable agility. Despite his age, Ma took an active role in farming tasks — plowing fields, fertilizing crops, and tending to the land, both in the sweltering heat of summer and the freezing cold of winter. He consistently performed these tasks himself, reflecting his deep dedication to the work. Each time the author, Lan Guangce, visited the garden, Ma personally guided him, attending to small tasks like opening doors, holding lanterns, and leading the way, without ever relying on servants. The garden was meticulously organized, with flourishing flowers, lush greenery, and thriving vegetables, reflecting Ma's expertise in agriculture. Lan Guangce, impressed by the garden's careful design and management, concluded that the Ma family's prosperity was a natural outcome of their hard work. In the inscription's closing, Lan highlighted that agriculture should take priority over military preparations, as only through farming could people achieve lasting prosperity — a core principle of Confucian governance. The inscription encapsulates Ma's philosophy of balancing intellectual pursuits with a deep dedication to working in the fields.

镜清堂
Jingqing Belvedere

As recorded in *The Prominent Families of Qingpu*, the Ma family upheld the ancestral principle of "valuing knowledge and reason over seeking official positions." This philosophy enabled the family to successfully manage their business across generations, amassing wealth and influence that benefited the local community. Meanwhile, the Ma family was also deeply involved in charitable work. Family records show they funded the construction of Diannan Elementary School, Zhonglong Bridge and Yongquan Bridge on West Jingting Port, as well as established charitable farmlands. Additionally, they contributed to public safety efforts to prevent banditry and supported many other community service initiatives.

In the northwestern corner of Kezhi Garden stand a stone carving titled *Zhuxi Five Elders* and an accompanying inscription stele. The carving depicts a 1925 gathering, where 73-year-old Ma Weiqi and local gentry members Cai Yiyu, Dai Zaoliang, Shen Hongxuan, and Jin Minchen met at Kezhi Garden to compose poetry and exchange literary works. The inscription, written by Zhang Ru of Huating, calligraphed by Pan Baoyi, and engraved by Zhou Meigu, is carved in elegant small regular script on bluestone. It records the contributions these esteemed elders made to the local community.

Ma Weiqi's approach to life is captured in the *Biography of the Farmer and Gardener Mr. Ma*. In April 1918, Lan Guangce, then serving

园景
Garden view

周梅谷将此文制成石碑,立于园中供人阅览。石碑原位于课植园内假山旁,为青石质地,高78厘米,宽163厘米,后断为三块。石碑虽断,然意绵长。

历经百年沧桑岁月,2012年,值课植园百年之际,这段历史经考证借残碑在园中得以重现。虽碑文已漫难辨,幸好《马氏家乘》中附有碑文。碑文开头引用《论语·樊迟请学稼》前半段,以此表达耕植工作必须要有切实的实践经验,非空有称号便能有成果。内容中提及,马家园主名为维骐,号文卿,昆山井亭乡人,七十多岁,鹤发童颜,不蓄长须,容貌似四五十岁之人,健步如飞。平素的耕植劳作中,马园主事必躬亲,夏季满身汗水污泥,冬天冒着严寒冰霜,赶牛犁地,为田地施肥,坚持亲力亲为。蓝光策每次游园时,马维骐都会亲自陪同,诸如开门、提灯、导引等琐事都不让仆人代劳。园子布置井然有序,花木繁茂,蔬菜丰茂,足以说明马园主极具农圃之经验。透过园内的规划经营,蓝光策认为马家富甲一乡乃必然。碑文之末,蓝光策还表达了自己的感悟。他认为农耕应先于军备,百姓方能生活富足,此为孔子经世之略。可以说,该碑文是马园主"课读之余,不忘耕植"理念的生动写照。

价值特色

Features and Significance

据《青浦望族》记载[2],马家历代传承"读书明理,不以仕途为官"的祖训,使马家世代家业经营有方,能够积累财富,赀雄乡里。同时,马家不忘积极参与社会公益事业,造福一方。据家谱记载,马家曾捐建了淀南小学大部分校舍及西井亭港的中龙桥和涌泉桥,购置义田,捐款防匪以维护治安,善举颇多。

课植园西北处,有《珠溪五老图》和序文碑刻各一块。《珠溪五老图》刻画了1925年,73岁的马维骐与地方乡绅蔡一隅、戴造梁、沈鸿轩、金敏臣会聚于课植园诗文唱和的情景。序文碑刻由华亭张汝撰写,潘宝沂书丹,周梅谷镌刻,为青石质地,小楷字体,文字俊秀,其内容记述了他们作为乡绅耆老为当地所作的贡献。

马维骐的处世之道还可见于《农圃者马君传》一文中。民国七年(1918)4月,时任县知事的蓝光策到井亭镇考察虫害情况,寓居于马维骐的课植园内。经过长期的交往和观察,蓝光策对马维骐的价值观念表示认同,于是书写了《农圃者马君传》一文,后来马维骐请

《珠溪五老图》碑刻
Twin steles of *Zhuxi Five Elders*

月圆地穴
Moon gate

课植园内曲水东西跨有一桥,名为"课植桥",桥长5米,为单跨石梁桥。以桥为界,东为课园,西为植园。根据原有照片记录,原本植园中有处"稻香村",面对稻香的位置建有"红房子",即"耕九余三堂",是一座典型的西式二层洋楼建筑。如果说藏书楼体现了课园的"课经书学千悟万",那原"红房子"就是植园"植稻麦耕九余三"的点睛之作。园主马维骐晚年时自称"农圃者",曾在园中种植瓜果、花草、水稻等,他培育出的各种名优品种的兰花,被人称作"马家兰花",驰名远近。他还亲率耕植,培育出了"青角薄稻米",成为当时朱家角的盛产品牌。"植园"疏朗开阔,别具人文意境和田埂野趣,更蕴涵了科学实证的思想内涵。可惜,曾经课植园的命途多舛,使如今的课植园以课园为主,植园的种种仅存于人们的记忆中了。

　　Kezhi Bridge, a 5-meter-long, single-span stone structure, spans the winding stream in Kezhi Garden, marking the boundary between the Ke Garden to the east and the Zhi Garden to the west. Historical photos show that Zhi Garden once featured an area called Daoxiangcun (Village of Fragrant Rice Paddies), with a Western-style two-story building known as the Red House overlooking the paddies, which is "Gengjiu Yusan Tang" (Hall of Nine Years' Toil and Three Years' Reserves). If the Library Tower in Ke Garden embodied the scholarly spirit, then the Red House in Zhi Garden was the crowning feature of its agricultural heritage. In his later years, Ma Weiqi referred to himself as a "farmer and gardener." He cultivated fruits, flowers, and rice, including prized orchids, famously known as "Ma Family Orchids" throughout the region. Ma also oversaw the cultivation of a prized thin-husk rice variety, renowned in Zhujiajiao for its high yield and exceptional quality. The Zhi Garden was spacious and open, blending cultural richness with rustic charm and featuring scientific agricultural experimentation. Sadly, over time, Kezhi Garden has experienced many hardships, and today, the garden primarily consists of the Ke section, while the Zhi section lives on only in the memories.

课植桥与稻香村
Kezhi Bridge and Daoxiangcun

上：课植桥
U Kezhi Bridge
下：课植桥侧景
D: Side view of Kezhi Bridge

从课植园厅堂区北侧，可进入一座中西合璧的"书城"，别称"藏书楼"。藏书楼共两层，飞檐翘脚，建筑风格为中西融合，楼房的柱、栏杆和楼前抱梯均采用水泥砖石。藏书楼南外墙贴着一座精巧的拱形旱桥，翠绿色琉璃瓦筒桥栏，水磨石子阶梯、竹节形扶手，寓意"节节高"。通过旱桥两侧可至该楼二层，拱形旱桥的巧妙之处在于利用拱形空间建成一个月形洞门，与左右立墙上的月形洞门相照应，游园者置身于天井中，会发现四周都有一个月形的洞门，此一物得二景，一显一隐，值得赏园者细品。此楼当时为"课读"之用，体现课植园的"课"之博大内涵。藏书楼前有座双帛亭，亭内有双井一对，旧时门前建井有招财之意，"帛"古时指丝绸，后引申出"和"的含义，遂有"化干戈为玉帛"的说法。故双帛井有聚拢财富、家和万事兴之寓意。

North of the hall area in Kezhi Garden lies a Sino-Western fusion structure known as the "Book City" or "Library Tower." This two-story building, with upturned eaves, blends Chinese and Western architectural elements. Its pillars, railings, and staircase are made from cement, bricks, and stone. A beautifully crafted arched dry bridge on the southern exterior wall features green-glazed tile railings, terrazzo steps, and bamboo-shaped handrails, symbolizing "steady progress." The bridge leads to the second floor and ingeniously incorporates a moon-shaped doorway beneath the arch, harmonizing with the moon gates on the side walls. As visitors explore the courtyard, they will notice moon-shaped doorways on all sides — some visible, others hidden — offering a delightful architectural feature. This tower was originally used for study, embodying the significance of "studying" in the garden's name. In front of the Library Tower stands Shuangbo Pavilion, which houses two wells. In ancient times, wells near the entrance were believed to attract wealth. The word "*bo*" in the pavilion's name traditionally referred to silk fabrics, often given as state gifts, and symbolized "turning conflict into peace." Thus, Shuangbo Pavilion represents both the gathering of wealth and the prosperity that comes from family harmony.

藏书楼与双帛亭

Library Tower and Shuangbo Pavilion

上：藏书楼
U: Library Tower
下：双帛亭
D: Shuangbo Pavilion

水晶宫
Shuijing Palace

  穿过带脚头单料月圆地穴，沿着水池可看到一座观鱼台，也称"水晶宫"，为2008年仿原貌重建。观鱼台朝东有一开门，一级级石阶伸向水面，小屋犹如浮在水中，鱼儿喜聚于此，园主马维骐闲暇时喜爱来此"数鱼"。

  Visitors pass through a moon gate and follow the pond's edge to the Fish Viewing Terrace, also known as the Shuijing Palace. Rebuilt in 2008 to resemble its original form, the terrace faces east, with stone steps descending into the water, making the small structure appear as if it floats on the surface. Here, fish gather in abundance, and Ma Weiqi often spent his leisure time "counting fish."

打唱台
Opera Stage

假山池塘近旁建有打唱台、观唱楼等建筑，戏台呈进深较浅的长方形，造型和陕西关中地区尚存的古戏台形制相似，从其风格不难看出，园主马维骐一直以身为东汉伏波将军马援的后裔为荣，借此楼台的形制表达对马氏远祖的追怀。观唱楼背靠水池，是品酒观月的佳地，别称"水月榭"。据说每逢佳节，马维骐会邀请镇上乡亲到园中一同看戏。

Near the rockeries and pond, you'll find the Opera Stage and the Viewing Tower. The stage, a narrow rectangular structure, resembles the ancient opera stages found in the Guanzhong region of Shaanxi Province. Its design reflects the pride that Ma Weiqi took in his heritage as a descendant of Ma Yuan, the renowned General Fubo of the Eastern Han dynasty. The stage stands as a tribute to his distant ancestor. The Viewing Tower, also known as Shuiyue Gazebo, sits by the pond and provides an ideal setting for enjoying wine under the moon. During festivals, Ma Weiqi would invite townspeople to the garden for performances.

以碑廊为界，沿着曲径行至南部假山区。该园的假山石多为精选的太湖石，重峦叠嶂，曲径通幽，经园主精心布局，成为蜿蜒曲折的"迷宫"。假山旁有池塘，池内种植荷花，仿上海豫园之池景。静中生趣，动静结合，为此处山石增添了几分雅趣。穿过假山，可见"倒挂狮子亭"，俗称"五角亭"。依古时建筑惯例，亭子通常为四角或六角。此亭飞翘五角，亭内倒挂五只造型生动、憨态可掬的狮子，造型别具一格，五个角对应中国的五行，寓意着五行相生、五谷丰登、五狮贺喜。倒挂狮子亭虽模仿他处，但不落因袭，有主有宾，互相"因借"。

Beyond the Stele Corridor, a winding path leads to the southern rockery area. The garden's rockeries, composed primarily of carefully selected Taihu rocks, create an intricate landscape of peaks and valleys. Designed with a labyrinthine effect, the winding paths and secluded corners were thoughtfully arranged by the garden's owner. Adjacent to the rockeries is a lotus-filled pond, modeled after the pond in Shanghai's Yu Garden. This harmonious combination of serene waterscape and dynamic rock formations adds a refined elegance, balancing tranquility with movement. Just beyond the rockeries stands the Daoshi Pavilion, also known as the Five Corner Pavilion. While traditional pavilions typically have four or six corners, this unique structure features five upturned corners, with five playful lions hanging upside down. The pavilion's five corners symbolize the Five Elements (metal, wood, water, fire, and earth), representing harmony. They also symbolize the Five Grains (rice, wheat, foxtail millet, proso millet, and soybeans), which signify a bountiful harvest. Meanwhile, the five lions represent joyful celebration. Different from traditional designs, the Daoshi Pavilion preserves its originality, blending tradition and creativity in a distinctive and harmonious way.

倒挂狮子亭外景
Exterior of Daoshi Pavilion

倒挂狮子亭内景
Interior of Daoshi Pavilion

倒挂狮子亭
Daoshi Pavilion

望月楼
Wangyue Tower

迎曛阁边建有一楼，名为"望月楼"，又名"冠云挚月楼"，这是一座五层方形建筑，水泥砖石砌筑，四向开窗，楼顶建有四角亭。旧时登楼远眺，可饱览淀山湖、大淀湖、漕港河三水胜景，课植园全景也可尽收眼底。望月楼是当时全镇最高的建筑，在湖匪猖獗时期曾发挥警戒周边的作用。

Adjacent to Yingxun Belvedere is Wangyue Tower, also known as the Guanyun Zhiyue Tower. This five-story square building, constructed with cement, bricks, and stone, has windows on all four sides and a pavilion at the top. In the past, climbing the tower offered sweeping views of Dianshan Lake, Dadian Lake, and Caogang River, as well as the entirety of Kezhi Garden. As the tallest building in town at the time, it also served as a lookout post during periods when lake pirates were prevalent.

迎曛阁
Yingxun Belvedere

迎曛阁外景
Exterior of Yingxun Belvedere

　　碑廊端接迎曛阁（原读书楼），阁为两层，可观夕阳西下之景，左为平顶碑廊，右为两层廊，建筑为砖石结构，属歇山顶形制，铺设欧洲进口花面地砖，四周均造走廊，走廊上用琉璃瓦筒作栏杆，采用中西结合的风格。

At the end of the corridor stands Yingxun Belvedere, formerly the Dushu Tower, a two-story structure offering views of the setting sun. To the left is the flat-roofed Stele Corridor, and to the right is a two-story corridor. Both are constructed from brick and stone, featuring hip-gable roofs and patterned floor tiles imported from Europe. The corridors are lined with glazed tile railings, showcasing a blend of Chinese and Western architectural styles.

在厅堂区的南边有一长廊，长廊纵向相隔，是清代建园常用的阴、阳复廊（古时女子走阴廊，男子走阳廊）。阳廊尽头是一段长约 20 米的碑廊。碑廊里共有 15 块碑刻，是园主马维骐费尽心思搜罗而来的明代才子书法墨宝，并请来有名金石匠刻碑建廊。据《青浦碑刻》记载，碑廊中的碑刻分别为：文徵明游西山诗十二首，共 2 块，字体为行草，清晰可辨；周天球诗二首，行草体，文字清晰；祝枝山梅花诗五首，共 11 块，狂草书体；唐寅书简，共 2 块，行书兼草体。这些碑刻都为青石质，至今仍保存完好，实属不易，现已外加玻璃木框罩之。园内的碑廊，不仅是一种借鉴，也体现了马家崇尚传统文化的高雅品格，使课植园里散发着浓郁的书卷气。

To the south of the hall area is a long corridor, divided into sections, known as the "Yin-Yang Corridor" — a traditional Qing dynasty feature where women would walk in the Yin Corridor and men in the Yang Corridor. At the end of the Yang Corridor lies a 20-meter-long Stele Corridor. This gallery contains 15 steles, carefully curated by the owner, Ma Weiqi, featuring calligraphy treasures from scholars of the Ming dynasty. Ma employed renowned stone craftsmen to carve and erect these steles. According to the *Qingpu Steles*, the gallery contains several remarkable works. Among them are two steles featuring 12 poems by Wen Zhengming about his travels to West Hill, written in cursive script with characters that remain legible. Additionally, two poems by Zhou Tianqiu are similarly inscribed in cursive and are well-preserved. Zhu Zhishan's five poems about plum blossoms are distributed across 11 steles in a wild cursive style, and two steles display a letter by Tang Yin, combining running and cursive script. These inscriptions are all made of greystone well-preserved and are now protected behind glass and wooden frames. The steles remain in remarkable condition, a testament to their historical value and careful preservation. The Stele Corridor in Kezhi Garden not only draws on traditional influences but also reflects the Ma family's deep respect for classical culture, imbuing the garden with a scholarly ambiance.

碑廊
Stele Corridor

上：碑廊外景
U: Exterior of Stele Corridor
下：阴阳廊
D: Yin-Yang Corridor

迎贵厅入口
Yinggui Hall entrance

后花厅内景
Interior view of Houhua Hall

Upon entering Kezhi Garden, the central axis leads through four successive halls: the Entrance Hall, Reception Hall, Houhua Hall, and Yinggui Hall. Each courtyard in front of these halls features intricately carved brick gate towers, with the ground paved in terrazzo tiles arranged in ornate patterns. The detailed brick carvings and floral decorations reflect the wealth and cultural sophistication of the garden's owner. The Entrance Hall, also known as the Sedan Hall, was designed as a resting place for sedan chairs. The second hall, the Reception Hall, was where the garden owner would host guests. It features large pillars, thick beams, and floor-to-ceiling windows with floral trellises. The beams are intricately carved and painted, and the exterior corridor is adorned with decorative eave panels and swastika-patterned railings. Inside, a prominent couplet reads: "Through studying the classics, one gains endless wisdom; Through tilling the fields, nine years of toil yield three years of reserves." Known as the "Kezhi Couplet," it reflects the garden's name, Kezhi — representing studying (Ke) and tilling (Zhi). It embodies Ma Weiqi's hopes for his descendants, encouraging them to pursue both education and agriculture to ensure lasting prosperity. The third hall is the Houhua Hall, also known as the "Tanglou Hall." It extends approximately 13 meters in depth, and its doorway features a brick plaque inscribed with "*Yonghuai kehu*" (Forever cherish and emulate noble ambition), a quote from the *Book of the Later Han*. The Ma family chose this inscription to encourage future generations to aspire to the virtues and character of noble individuals. The fourth hall is Yinggui Hall, designated for receiving honored guests and hosting important discussions. In front of the hall, osmanthus trees are planted, symbolizing "welcoming honored guests." This hall is particularly exquisite, with intricately carved beams and finely painted rafters. The roof's interior features a double-layer tile structure. The flooring consists of beautifully patterned tiles imported from Europe. These high-quality tiles have withstood the test of time, remaining intact with vivid patterns.

迎贵厅内景
Interior of Yinggui Hall

迎贵厅地砖
European-imported floor tiles in Yingui Hall

进入园区，沿中轴布置有四进厅堂，分别是门厅、会客厅、后花厅、迎贵厅。厅堂前的天井有砖雕门楼，地面用印纹水磨石拼出各式图案，精细繁复的砖雕花饰显示出主人家的富贵和文化素养。墙门内是门厅也是轿厅，专为停歇轿子所用。第二进是会客厅，是园主会见客人的地方。会客厅柱大梁粗，花架落地长窗，画栋雕梁，图案精细。外侧弯橡走廊，走廊外侧上饰檐口花板和天落，下饰"卍"字栏杆，厅内正面悬有一副"课经书学千悟万，植稻麦耕九余三"的对联，称"课植联"，此联既映衬了园林之名——课植园，又寄托了园主马维骐的厚望。这是马维骐留给子孙后代的"家训"，意为耕植庄稼要勤恳踏实，稼穑丰稔，吃用有余。第三进为后花厅，俗称"堂楼"，进深约 13 米，后花厅（堂楼）硬山式八字垛头墙门上砖雕匾额刻"永怀刻鹄"，语出《后汉书·马援传》，马家以此来教育后代，要永远牢记向品德高尚的人学习。第四进为迎贵厅，是接待贵客议事之处。厅前植桂花，取其谐音"迎贵"之意。此厅堂十分考究，雕梁画栋，工艺精细，屋顶内面采用双层瓦片行板结构。地面贴砖属西式特点，是园主从欧洲进口的花纹地砖，地砖图案精美，质量上乘，历经多年，毫无损坏，图案依旧清晰。

会客厅
Reception Hall

后花厅（堂楼）硬山式八字垛头墙门
Houhua Hall and Brick Plaque Inscription

四进厅堂
Four Successive Halls

会客厅入口
Entrance to the Reception Hall

园区游客入口
Visitor entrance of the park

园景撷趣
Garden Scenery

　　课植园的布局特点突出表现了传统造园类型中宅第园林"园居"的形式。课植园西临井亭港,原有五门。门前有船舫和港埠等泊船设施。大门前原有一垛数米高的照墙,两旁曾建有两座"司鼓亭"。试想每逢佳节,曾经的"马家花园"必是鼓乐齐鸣,门庭若市,贵宾络绎不绝,园主热情相迎,一派家业昌盛之景。

Kezhi Garden's layout exemplifies the "garden residence" style typical of traditional private estate gardens. Located along Jingting Port to the west, the garden originally had five gates. In front of the main gate were docking facilities, including piers and moorings for boats. A tall screen wall once stood before the entrance, flanked by two Drum Pavilions. One can imagine the lively atmosphere of the Ma Family Garden in its prime, filled with the sounds of drums and music, bustling with guests, and the garden owner warmly welcoming distinguished visitors — a true reflection of the family's prosperity.

In 1912, Ma Weiqi (1853–1928), a member of the Ma clan in Zhujiajiao, began constructing an estate at the northern end of the ancient town, now located on Xijing Street. Known as Kezhi Garden, or "Ma Family Garden," the estate originally covered 96 *mu* (64 000 square meters, existing 17 997 square meters) and cost around 300 000 taels of silver. Key structures in the garden included the Tanglou Hall, Yinggui Hall, Library Tower, Wangyue Tower, Daoshi Pavilion, Opera Stage, Stele Corridor, and others. Ma Weiqi, courtesy name Lingyuan and literary name Wenqing, was born during the Xianfeng period (1851–1861) of the Qing dynasty. As a youth, he was sent by his father to Suzhou to learn the trade at an old-style private bank and later expanded the family's salt business. Ma eventually diversified into various forms of foreign trade, with the export of Jiangnan silk proving especially lucrative. Embracing Western culture early on, Ma incorporated elements of it into the design of Kezhi Garden, creating a fusion of Chinese and Western styles that made it the largest private garden in the area. Construction of the "Ma Family Garden" took 15 years, and although it was not fully completed by the time of Ma Weiqi's death, the garden was already quite grand. It is said that Ma traveled extensively through the famous gardens of Jiangnan, drawing inspiration from each. Whenever he encountered a particularly beautiful scene, he had it replicated in his own garden. Kezhi Garden reflects Ma's love of travel and his desire for a peaceful life of retreat.

After its completion, Kezhi Garden suffered significant damage during several wars, with structures like the Shuijing Palace and the Baifu Pavilion being destroyed, and the garden's rice paddies falling into neglect. In 1956, the garden was repurposed as Zhujiajiao Middle School, and in 1986, it was designated a protected historical and cultural site in Qingpu County. Restoration work began in 1999, and the garden officially reopened to the public in October of that year. To preserve the garden's historical architecture, the local government reconstructed the Shuijing Palace and Baifu Pavilion, restored the Nine-Bend Bridge, Yin-Yang Corridor, and Yinggui Hall, and reinforced the corridors of the Shuiyue Gazebo. Structural reinforcements and safety inspections of the ancient buildings were also conducted. A plaque inscribed with "Kezhi Yuan" (Garden of Studying and Tilling), written by renowned calligrapher Xu Boqing, was added to the entrance wall, welcoming visitors to the garden in its fully restored state.

望月楼远景
Distant view of Wangyue Tower

打唱台侧景
Side view of Opera Stage

课植园大门
The gate of Kezhi Garden

说,为造这座花园,马维骐游览了多处江南名园,博采众园林之长,每见一处胜景,就命人着意仿建至园中,可以说,这座江南园林蕴含了园主对游居生活的钟爱和对退隐息躬的田居生活的向往。

课植园建成后,多次经历战火,受损严重,园内水晶宫、百蝠亭等建筑皆遭损毁,植园稻田长久荒废。1956年始,课植园被辟为朱家角中学,1986年,课植园被公布为青浦县文物保护单位。1999年起逐步予以修复,1999年10月正式对外开放。在此期间,为实现园内历史建筑的保护与传承,当地政府先后复建了水晶宫、百蝠亭,修缮了九曲桥、阴阳廊、迎贵厅,对水月榭连廊实施加固,加强了对课植园古建筑的安全检查与监测工作,还在门墙上镶嵌了一幅由徐伯清先生书写的"课植园"铭牌,以使课植园能以如今饱满的姿态迎接各方游客。

历史沿革

Garden History

民国元年(1912),朱家角井亭马氏族人马维骐(1853—1928)在朱家角古镇北端(今西井街处)开始建造一座庄园,称作"课植园",别称"马家花园"。课植园坐西朝东,当时占地96亩(64 000平方米,现存17 997平方米),耗资约30万银两。园内主要建筑有堂楼、迎贵厅、藏书楼、望月楼、倒挂狮子亭、打唱台、碑廊等。园主马维骐,字龄苑,号文卿,国学生,生于清咸丰三年(1853),少年时便被父亲送去苏州钱庄学做

迎贵厅内景
Interior of Yinggui Hall

入口铭牌
Entrance plaque

生意,青年时随父学习经营盐业,后来生意不断扩大,涉及外贸等诸多种类,尤以出口江南精品丝绸获利最丰。他对西洋文化接受较早,因而在园林建筑、装饰设计方面,将中国建筑艺术与西洋建筑文化加以融合,洋为中用,把园林建成为中西合璧的庄园式私家花园,其规模在当地属最大。一座"马家花园",建园耗时15年,直至马维骐离世,仍未全部建成,但已初具规模。据

清泉石涧
Clear spring and rocky stream

中国古典园林被誉为中国文化"四绝"之一,它的营造多以宅建、山石、水和植物为构成元素。人居于宅,是人伦亲情的体现;人寓于园,是追求精神的无拘无束,宅与园的结合成为传统生活的载体。富贾文人皆偏爱私家园林,诸多精华建于江南一隅,现位于上海市青浦区朱家角镇的课植园便是其中之一。

千年朱家角,百年课植园。据《朱家角乡志》记载[1],朱家角旧称"珠里",早在5000年之前便有先民在此繁衍生息。朱家角镇被太湖水系的淀山湖、大淀湖、漕港河环抱,境内河港纵横,土地肥沃,曾有"三泾不如一角"之说,是典型的江南鱼米之乡,课植园就位于这座千年古镇的北端。

Chinese classical gardens are one of the "Four Great Accomplishments" of Chinese culture, featuring key elements such as residences, rockeries, waterscapes, and plants. The residence represents family life, while the garden provides a retreat for spiritual reflection. Together, they capture the essence of traditional Chinese living. Wealthy merchants and scholars were particularly fond of private gardens, with many of the finest examples concentrated in the Jiangnan region. Kezhi Garden in Zhujiajiao Town, Qingpu District is a remarkable example of this tradition.

As detailed in the *Gazetteer of Zhujiajiao Township*, Zhujiajiao, formerly known as "Zhuli," has been inhabited for over 5000 years. Nestled between Dianshan Lake, Dadian Lake, and the Caogang River — part of the Taihu Lake water system — the town's fertile landscape is interlaced with rivers and canals, making it renowned for its agricultural abundance. Kezhi Garden sits at the northern edge of this historic town.

课植园
Kezhi Garden

课经植道，积厚流光——课植园
Cultivating Virtue and Legacy: Kezhi Garden

年代：清代
地址：上海市青浦区朱家角镇西井街119号
占地面积：17 997 平方米
保护级别：青浦区文物保护单位

Era: Qing dynasty
Address: No.119, Xijing Street, Zhujiajiao Town, Qingpu District, Shanghai
Area: 17 997 square meters
Protection level: Qingpu district-level protected site

课植园鸟瞰图
Aerial View of Kezhi Garden

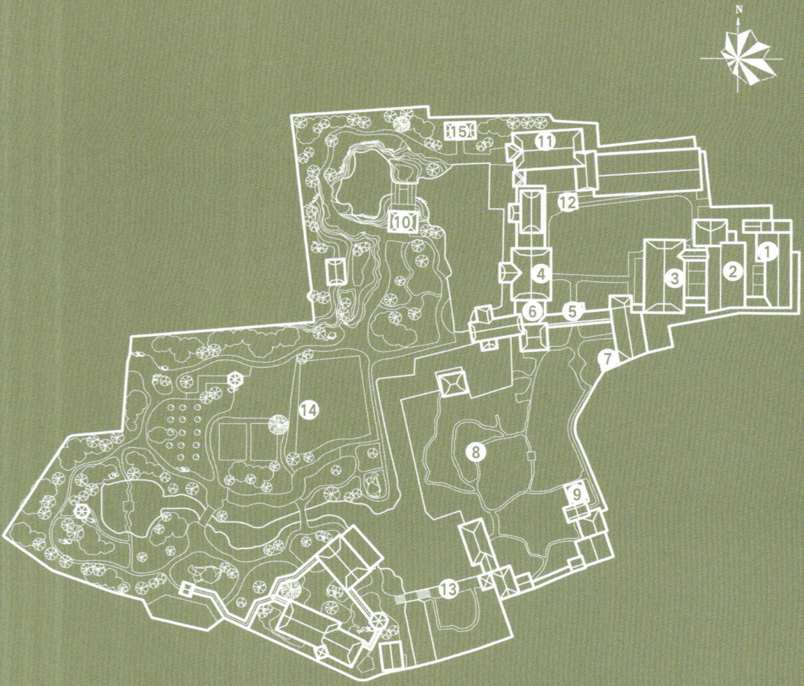

图例
Legends

1 门厅
　Entrance Hall

2 会客厅
　Reception Hall

3 后花厅
　Houhua Hall

4 迎贵厅
　Yinggui Hall

5 碑廊
　Stele Corridor

6 迎曛阁
　Yingxun Belvedere

7 望月楼
　Wangyue Tower

8 倒挂狮子亭
　Daoshi Pavilion

9 打唱台
　Opera Stage

10 水晶宫
　Shuijing Palace

11 藏书楼
　Library Tower

12 双帛亭
　Shuangbo Pavilion

13 课植桥
　Kezhi Bridge

14 稻香村
　Daoxiangcun

15 《珠溪五老图》
　Zhuxi Five Elders

课植园

Kezhi Garden

曲水园秋景
Autumn scene in Qushui Garden

多种名贵品种也会集中展示，引得无数游人争相赏秋。

2007年，曲水园被评为上海市四星级公园，这座清代古园以其独特的风景吸引着越来越多游客驻足。漫步于此，处处静谧秀雅，令人流连忘返。

曲水园中的菊花展
Chrysanthemum exhibition in Qushui Garden

Among Shanghai's 100-plus parks, Qushui Garden holds the largest and most concentrated collection of ancient and renowned trees, celebrated for its "cherry blossoms in spring, lotus blooms in summer, fragrant osmanthus in autumn, and radiant wintersweet in winter." Upon entering, visitors are welcomed by twisting wisteria, graceful pines and cypresses, elegant camellias, and towering ginkgo trees. At the entrance, bonsai arrangements of pine, plum, and rocks lend an antique charm, creating a lush landscape filled with vibrant blossoms and abundant fruit.

Qushui Garden is home to 53 ancient and notable trees, which blend harmoniously with its architecture, rockery, and waterscape. Near Jiejing Corridor in the eastern garden stands a wisteria vine dating back to the Qianlong period (1736–1795), now over 200 years old, blooming in early summer like a canopy of purple clouds. On Xiaofeilai Rockery, four cypresses over 200 years old and three maples over a century old form a dense, leafy canopy, shading the Jiufeng Yilan Tower atop the rockery. From a distance, only the tower's rooftop is visible, adding a sense of majesty and mystery.

In winter, visitors enjoy the plum blossoms; in summer, the lotuses; in spring, the peonies; and in autumn, the chrysanthemums. Among these seasonal blooms, peonies and chrysanthemums are the most celebrated in Qushui Garden. The garden boasts around 200 peony plants of various types, filling the grounds with vibrant colors each spring. The annual autumn chrysanthemum exhibition, a tradition since the Republic of China period, features nearly ten thousand potted chrysanthemums. Rare varieties are showcased in dedicated areas, drawing crowds to experience the beauty of autumn.

In 2007, Qushui Garden was designated a four-star park in Shanghai. This historic garden, with its distinctive scenery, continues to captivate a growing number of visitors. As they explore its serene, graceful paths, visitors often find themselves staying longer, immersed in the garden's peaceful charm.

古木芳菲

Flowers and Trees

春天的曲水园（上海曲水园提供）
Qushui Garden in Spring (Provided by Shanghai Qushui Garden)

在上海的100多座公园中，曲水园古树名木最多、最集中，素有"春日樱桃争艳，夏天荷花出水，入秋金桂馥郁，冬令蜡梅璀璨"之誉。走进园林，就见紫藤盘绕、松槐婆娑、山茶多姿、银杏参天，门口松梅山石盆景古朴典雅，好一派枝荣叶茂、花繁果硕的美景。

曲水园中仅挂牌的古树名木就有53棵，它们与亭台楼阁、池塘古桥掩映成趣、浑然一体。园中东面接景廊北侧有1棵栽于乾隆年间（1736—1795）的紫藤，已有200多岁，每年初夏依然花开繁盛，宛如漫天紫霞。小飞来峰上有4棵超过200岁的桧柏，有3棵树龄百年以上的青枫，几棵古树冠宛如华盖，将假山顶上居高临下的九峰一览阁掩映在浓荫翠绿之中，远观只露出一顶，增添了几分威严和神秘之感。

冬赏梅花夏赏荷，春有牡丹秋有菊。牡丹和菊花是曲水园里四时花卉中最为著名的。园内有200多株牡丹，品种繁多，每年春季，满园富丽。每年秋季的菊花展是曲水园自民国以来的传统花卉园艺展示项目，近万盆菊花布置在园中各处，

参观指南 Visiting Guide

开放时间：06:00—21:00（20:30 停止入园）。
交通信息：地铁17号线至漕盈路站，换乘青浦9路至公园路城中北路站。

Opening hours: 06:00-21:00 (20:30 Last admission).
Traffic information: Metro Line 17 to Caoying Road Station, transfer to Qingpu Bus 9 at Gongyuan Road (North Chengzhong Road) Station.

御书楼
Yushu Tower

After weathering the passage of time, Qushui Garden has reached a refined state and now offers a variety of cultural and recreational activities, making it a favored spot for residents of Qingpu's old city to relax and enjoy entertainment.

Since the Republic of China period, the garden has hosted calligraphy and painting exhibitions, including notable shows like Qian Tongfu's calligraphy exhibition from Wujin District and Tang Yifang's painting exhibition from Songjiang District, both cherished in local history. This tradition continues today. In May 1989, Qingpu artist Cen Zhenping established the "Qingpu Art Gallery" in Deyue Gallery, with its name inscribed by the renowned artist Cheng Shifa. The gallery regularly features local artists' work, enriching the cultural life of the community.

Spring and autumn outings have become cherished traditions in the garden. Established during the Republic of China period, the Xintuanyue Teahouse still operates within the garden today. As the seasons turn, visitors can savor tea in this serene setting, evoking scenes of "a gentle midday breeze, trees and flowers in full summer bloom, the scent of lotus wafting over the water, and cicadas singing from every hillside."

In addition, the Qushui Garden is adorned with ancient stone drum inscriptions, plaques, couplets, poems, and calligraphy, creating a visual feast. During the 2005–2007 restoration, the garden department commissioned 30 new couplets for various scenic spots and added over 20 new plaques. With support from Fudan University's Chinese Literati Calligraphy and Stone Drum Research Center, 10 replicas of pre-Qin stone drum inscriptions were placed in Guanshou Garden, while over 80 calligraphy inscriptions were embedded in the walls of Qingquan Corridor, earning it the title "Stone Drum Calligraphy Garden."

人文撷趣

Cultural Highlights

历经沧桑兴废，如今的曲水园渐趋完善，各种文娱活动不断，成为青浦老城厢里百姓休闲、娱乐的极佳场所。

曲水园自民国时期便举办书画展，武进钱彤夫书法展、松江汤义方画展等均文史留名。此传统一直延续至今。青浦画家岑振平于1989年5月在得月轩创办"青浦美术界画廊"，由著名书画家程十发题名。该画廊常年为本地书画家举办展览，持续为地方文化的繁荣贡献力量。

园中探春赏秋也蔚然成风。民国时期，园子里就设有新团月茶社，现在茶室依旧。寒来暑往，品茗其间，可谓"茗谈小憩午风轻，夏日炎炎花木荣。一阵荷香来水际，山前山后乱蝉鸣"。

此外，曲水园里的石鼓碑刻古朴厚重，匾额、楹联、诗文书艺令人目不暇接。2005—2007年园林整修期间，园林部门组织人员为园内的景点撰写了30副楹联，补写和新增了20余块匾额。在复旦大学中国文人书法暨石鼓文研究中心的帮助下，在观寿园中新添了10个仿制的先秦石鼓文石，清泉廊壁内镶嵌了80多幅书法碑刻，称其为"石鼓书艺苑"。

石鼓文书艺苑
Stone Drum Calligraphy Garden

牡丹亭建于1969年，亭为单檐六角攒尖形状，六个立面变化丰富，西面以门洞与接景廊相连，六边形洞窗上有牡丹亭匾额。绿波廊临河而建，蜿蜒曲折，起伏错落，古朴秀雅。从老东门进入青浦城，走上城河桥头，未入曲水园，先见的便是绿波长廊。

Peony Pavilion, built in 1969, features a single-eave, hexagonal design with a pyramidal roof and six distinct sides. On its west side, an arched doorway connects it to Jiejing Corridor, while a hexagonal window above displays a plaque with the inscription "Mudan Ting" (Peony Pavilion). Lübo Corridor, built along the river, winds gracefully with an antique charm. As visitors approach Qushui Garden from Qingpu's old east gate and cross the bridge over the city moat, this elegant, meandering corridor is the first feature to greet them.

绿波廊
Lübo Corridor

牡丹亭与绿波廊
Peony Pavilion and Lübo Corridor

牡丹亭
Peony Pavilion

机云亭
Jiyun Pavilion

  机云亭为曲水园众多亭子中唯一用人名命名的亭子。机云，指的是西晋著名文学家陆机、陆云两兄弟。二陆的祖父陆逊、父亲陆抗皆为东吴重臣。魏灭吴后，兄弟俩退居故里，闭门苦读十年，足见其节操。

  Jiyun Pavilion is the only pavilion in Qushui Garden named after historical figures. Ji and Yun honor the brothers Lu Ji and Lu Yun, renowned literary figures from ancient China's Western Jin dynasty. Their grandfather, Lu Xun, and father, Lu Kang, served as officials in the state of Eastern Wu, one of the Three Kingdoms of China. After Eastern Wu was conquered, the brothers returned to their hometown and devoted themselves to study for ten years, embodying values of dedication and scholarly integrity.

听橹阁与邀月廊
Tinglu Belvedere and Yaoyue Corridor

东园内,建于 2006 年的听橹阁,位于东园护城河畔,和旧时的码头隔河相望,置身此阁,仿佛能闻咿呀橹声。听橹阁南面为邀月廊,取自李白诗句中的"举杯邀明月",与南面的水月亭、牡丹亭、放生池、芳草地合成一院。

In the East Garden, Tinglu Belvedere, built in 2006, stands along the moat, directly across from the old pier. The name *tinglu* means "listening to oars," capturing the feeling of hearing the gentle sound of oars gliding through the water as visitors sit within the pavilion. To the south of Tinglu Belvedere is Yaoyue Corridor, whose name *yaoyue* means "inviting the moon," inspired by a verse from the famous Tang dynasty poet Li Bai: "I raise my cup to invite the bright moon." Together with Shuiyue Pavilion, Peony Pavilion, Free Life Pond, and Fragrant Meadow, Yaoyue Corridor creates a harmonious courtyard in the garden's southern area.

佛谷亭与小濠梁
Fogu Pavilion and Xiaohaoliang Pavilion

左：佛谷亭
L: Fogu Pavilion
右：小濠梁
R: Xiaohaoliang Pavilion

佛谷亭建在小飞来峰上，是一座内涵丰富的三角亭。小濠梁位于荷花池边，与恍对飞来、凝和堂、涌翠亭隔池相望，依山傍水，既可衬山高耸，又能相映成趣。

Fogu Pavilion, perched on Xiaofeilai Rockery, is a triangular pavilion imbued with symbolic meaning. Xiaohaoliang Pavilion, situated by the Lotus Pond, faces Huangdui Feilai Hall, Ninghe Hall, and Yongcui Pavilion. Nestled between rockery and water, it complements the surrounding peaks and creates charming reflections in the pond, enhancing the harmony of the landscape.

九峰一览阁
Jiufeng Yilan Tower

位于园林中央的小飞来峰,又称"大假山",山顶上的九峰一览阁是曲水园最高点。1927年,邑绅张景周捐款修假山,并建三层高台,时称"景周亭"。1937年淞沪会战时,公园大部分被日机炸毁。1949年后重筑大假山,山上恢复景周亭,后改名为"九峰一览阁"。登于阁上,当时可远眺北干山、横山、薛山、小砚山、辰山、佘山、天马山、凤凰山、小昆山九峰,故而得名。

Xiaofeilai Rockery stands at the center of Qushui Garden, crowned by Jiufeng Yilan Tower, which is the garden's highest point. In 1927, local philanthropist Zhang Jingzhou funded the construction of the rockery and a three-story structure known then as Jingzhou Pavilion. During the Battle of Shanghai in 1937, much of the park was destroyed by Japanese bombing. After 1949, Xiaofeilai Rockery was rebuilt, and Jingzhou Pavilion was restored and later renamed Jiufeng Yilan Tower. The tower's name, meaning "Nine Peaks in View," reflects the original view, once allowing visitors to see nine surrounding mountain peaks.

荷花池与喜雨桥
Lotus Pond and Xiyu Bridge

荷花池
Lotus Pond

  荷花池位于园林中央，是连接曲水园西园、中园、东园三大景区的枢纽，背靠大假山，面对凝和堂。喜雨桥，造于清乾隆年间（1736—1795），以花岗岩建成，构造齐全，精雕细凿。立在桥上观荷赏鱼，园中幽景尽收眼底。二桥为六曲桥和三曲桥，合则为九曲桥，被人们比作三国里的"大乔"和"小乔"两位美女，秀美脱俗，曲径通幽。

  The Lotus Pond sits at the heart of Qushui Garden, serving as a central hub that links the garden's western, central, and eastern sections. Backed by the grand rockery Xiaofeilai and facing Ninghe Hall, it provides a scenic focal point in the landscape. Xiyu Bridge, constructed during the Qianlong period (1736–1795) of the Qing dynasty, is made of granite and showcases exquisite carvings and craftsmanship. From the bridge, visitors can admire the lotus blossoms and fish, with serene garden views extending around them. Two additional bridges — the Six-Bend Bridge and Three-Bend Bridge — combine to form the Nine-Bend Bridge, celebrated for its elegance and winding paths that lead to tranquil retreats.

有觉堂
Youjue Hall

有觉堂始建于清乾隆十年（1745），于清光绪十五年（1889）重修，是曲水园中历史最悠久的建筑物，与凝和堂、花神堂并称为曲水园东西轴向线上的三大古建筑。有觉堂的四周有回廊环抱，南北开设刻花落地长窗，东西设装饰着大块玻璃的方形透视窗，从堂中可朝四个方向赏景，所以俗称"四面厅"。

Originally constructed in 1745 during the Qianlong period (1736–1795) of the Qing dynasty and renovated in 1889 during the Guangxu period (1875–1908) of the Qing dynasty, Youjue Hall is the oldest building in Qushui Garden. Together with Ninghe Hall and Flower God Hall, it forms part of the three historic structures along the garden's east-west axis. Surrounding Youjue Hall is a cloister, featuring floor-to-ceiling carved windows on the north and south sides and large, glass-paneled square windows on the east and west. These windows allow views in all four directions, earning it the nickname "Four-Sided Hall."

西园内,夕阳红半楼建在用太湖石垒成的假山上,楼下是盘旋曲折的山洞,楼的南北两侧均有石梯,在此欣赏晚霞漫天,可尽览怡然自得的诗情画意。舟居非水,俗称"旱舫""岸舫",建于清乾隆三十二年(1767),位于西园最南面,东傍南北贯通的小溪,西靠清澈的水池,北连得月轩,向西可至迎仙阁和夕阳红半楼。得月轩取"近水楼台先得月"之意,坐西朝东,全长60余米,为中国传统风格的砖木结构建筑。

In the garden's western area, the Xiyang Hongbanlou Tower sits atop a rockery made of Taihu rocks, with a winding cave beneath and stone steps on both the north and south sides. This spot offers an ideal view of the evening skies, evoking a sense of poetic beauty. Zhouju Feishui Stone Boat, often called the "dry boat" or "shore boat," was also built in 1767 during the Qianlong period (1736–1795) of the Qing dynasty. Positioned at the southern end of the West Garden, it sits beside a stream flowing north-south, with a clear pond on its western side. To the north, it connects to Deyue Gallery, while to the west, it leads to Yingxian Belvedere and Xiyang Hongbanlou Tower. Deyue Gallery, which translates to "Gallery of the First to Catch the Moon," is named after the saying, "The building by the water catches the moonlight first." Facing east, the gallery stretches over 60 meters and showcases a traditional Chinese brick and wood design.

得月轩
Deyue Gallery

夕阳红半楼与得月轩
Xiyang Hongbanlou Tower and Deyue Gallery

夕阳红半楼
Xiyang Hongbanlou Tower

步入曲水园,迎面可见淡灰色的砖雕门楼——仪门。此门建于清代,"文革"中被毁,2004年重建。门口有石狮一对,肃穆庄重。两边各有粉墙一堵,黛瓦盖顶,古色古香。凝和堂,宏丽轩敞,建于清乾隆三十二年(1767),为单檐歇山式,青瓦白墙,四方而高,庄严肃穆。堂脊饰以龙凤,象征福祥。

Entering Qushui Garden, visitors first encounter the light-gray brick-carved entrance archway, known as Yi Gate. Originally built during the Qing dynasty, the gate was destroyed during the Cultural Revolution and rebuilt in 2004. A pair of stone lions stand solemnly by the entrance, adding a dignified touch. Flanking the entrance, walls topped with gray tiles enhance the garden's traditional charm. Ninghe Hall, grand and spacious, was constructed in 1767 during the Qianlong period (1736–1795). Featuring a single-eave hip-gable roof, gray tiles, and white walls, this tall, square structure exudes a solemn dignity. The roof ridge is adorned with dragons and phoenixes, symbols of blessings and harmony.

凝和堂内景
Interior of Ninghe Hall

仪门与凝和堂
Yi Gate and Ninghe Hall

仪门
Yi Gate

园景撷趣
Garden Scenery

花神堂
Flower God Hall

　　曲水园坐北朝南，园内建筑多为青瓦、白墙、青砖，以凝和堂为中心，有觉堂、花神堂左右并峙，形成古典园林中较为少见的横向一轴三堂格局。三堂垣墙相隔，曲径相连，景色秀丽。园景布局以湖区为中心，环湖而增景，堂堂近水，亭亭靠池，大假山小飞来峰伴着荷花池、睡莲池，以山架二池水而闻名，使观者游园必绕池，故有"深深院落重重水，庭院深深深几许"之感。

Qushui Garden faces south, with most structures featuring gray tiles, white walls, and gray brickwork. At its center is Ninghe Hall, flanked by Youjue Hall and Flower God Hall, forming a unique horizontal axis with three interconnected halls — a rare arrangement in classical Chinese gardens. The halls are separated by walls yet connected by winding paths, creating a picturesque setting. The layout centers on water, with scenic spots arranged around; halls are positioned close to the water, and pavilions overlook the ponds. A grand rockery called "Xiaofeilai" complements ponds filled with lotus and water lilies. This design, featuring two ponds framed by a rockery, guides visitors around the ponds, evoking a sense of "courtyards unfolding in-depth, waters layered in mystery."

At this point, the garden featured a collection of 24 scenic spots, including Ninghe Hall, Xiyang Hongbanlou Tower, Huanbi Tower, Yingxi Pavilion, Yuzi Corridor, two bridges, Youjue Hall, Zhouju Feishui Stone Boat, Deyue Gallery, Jingxin Hut, Poxian Belvedere, Zhuojin Jetty, Flower God Hall, Huangdui Feilai Hall, Xiaohaoliang Pavilion, Xiyu Bridge, Mibai Pavilion, Binghu, Qingxu Jingtai, Tianguang Yunying, Yingxian Belvedere, Qiulong Cave, Baiyun Dock, and Qinglai Studio. With these restorations, the garden regained its former beauty, becoming the ideal spot in Qingpu for relaxation and leisure.

In 1911, after the Xinhai Revolution and the liberation of Qingpu, the City God Temple ceased its religious rituals, and Lingyuan Garden was transformed into a public park. From that time, Qushui Garden became a popular destination for sightseeing, relaxation, and social gatherings. In March 1926, to honor Sun Yat-sen, a leading figure in the revolution and often called the "Father of Modern China," Qushui Garden was renamed Zhongshan Park (Zhongshan being one of Sun's honorific names). Ninghe Hall, one of the garden's historic structures, was repurposed as the Zhongshan Memorial Hall to commemorate his contributions.

After 1949, Qushui Garden retained the name "Zhongshan Park" until 1980, when its original name was restored. Between 1984 and 1986, the Shanghai Municipal People's Government allocated 930 000 yuan for a comprehensive renovation, with Professor Chen Congzhou from Tongji University contributing plaques for scenic spots like Zhouju Feishui Stone Boat, Yingxian Belvedere, Xiyang Hongbanlou Tower, Youjue Hall, Yushu Tower, Deyue Gallery, and Poxian Belvedere. Over the next two decades, Qushui Garden underwent several rounds of restoration, the most extensive from 2005 to 2007, supported by 15 million yuan from municipal and district funds. This work included restoring structures, refining the layout, adding scenic spots, enhancing landscaping, and improving both functionality and aesthetics. Ultimately, 21 of the garden's 24 historical scenic spots — excluding Baiyun Dock, Mibai Pavilion, and Qingxu Jingtai — were preserved and restored, with updates that respect its historical heritage and refine the layout.

迎曦亭
Yingxi Pavilion

睡莲池
Water-lily Pond

scene reminded him of the "meandering stream" (*qushui*), described by the renowned calligrapher Wang Xizhi in his *Preface to the Poems Composed at the Orchid Pavilion*. Inspired, Liu renamed the garden "Qushui Garden" (Garden of Meandering Stream), a name that has endured to this day.

In 1860, during the Xianfeng period (1851–1861) of the Qing dynasty, Qingpu was occupied by Taiping forces. Two years later, in 1862, Qing troops and the Shanghai Volunteer Corps — a small imperial force trained in European tactics — joined forces to retake the city, and Qushui Garden was destroyed in the resulting shelling. Reconstruction began in 1884, during the Guangxu period (1875–1908) of the Qing dynasty, starting with the addition of Zhouju Feishui Stone Boat, Xiyang Hongbanlou Tower, and Xiaozi Hall. In 1887, the Yushu Tower was added, followed by Youjue Hall and Deyue Gallery in 1889, and a perimeter wall and main gate in 1890. By 1892, the garden saw the completion of Poxian Belvedere (originally built as a pavilion), Yuzi Pavilion, Yongcui Pavilion, and Xiyu Bridge, with land acquired east of the garden to create the Free Life Pond. Huangdui Feilai Hall was built in 1894, Ninghe Hall in 1897, and the Flower God Shrine and Binghu Pond in 1898. The final addition, Jingxin Hut, was completed in 1910 during the Xuantong period (1909–1911) of the Qing dynasty.

Throughout its long history, Qushui Garden has endured wars, name changes, and cycles of damage and reconstruction, yet it has steadily expanded eastward and northward. Two major renovations after 1949, in particular, have revitalized the garden's appearance.

Originally built in 1745 during the Qianlong period (1736–1795) of the Qing dynasty, Qushui Garden began as an annex to the Qingpu City God Temple. It has endured through seven Qing-dynasty reigns — from Qianlong to Xuantong (1736–1911) — and witnessed the Republic of China period, the establishment of the People's Republic of China (1949), and the reform and opening-up period, bearing witness to the transformations of each era.

Qingpu's City God Temple, originally built during the Ming dynasty's Wanli period (1573–1620), was once a vibrant center of worship. According to legend, in the spring of 1734, during the Yongzheng period (1723–1735) of the Qing dynasty, a monk traveling from afar arrived in Qingpu. Observing celestial signs, he predicted the end of the Yongzheng's reign and the rise of a new emperor. Sharing his insight with the temple's head, they pooled donations from temple patrons and the monk's savings to create Lingyuan Garden for the City God Temple. In 1736 during the Qianlong period, they began selecting a site, planning, and drafting blueprints. Over the next nine years, they completed the main layout of Lingyuan Garden, establishing structures like Youjue Hall and Deyue Gallery by 1745. By 1767, they had added Ninghe Hall and the Flower God Hall and dug canals. In 1784, they expanded the garden with ponds, rockeries, and plantings, and added pavilions and bridges. However, as the Qianlong period neared its end, the Triangular Pavilion in Lingyuan Garden was completed with only a simple thatched roof. Shortly after, the Qianlong period came to a close, and the monk's life also ended. While the details of the legend are difficult to confirm, records show that Qingpu's City God Temple received donations for repairs under a program called the "One-*Wen* Pledge." The *wen*, one of the main units of currency in historic China, was used to denominate coins. At that time, the temple and garden were managed together, and the garden's upkeep was funded by this pledge, leading locals to refer to it as the "One-*Wen* Garden." Another local story recounts that, around 40 years after Qushui Garden's completion, the Qingpu gentry proposed raising funds for the garden's expansion by adding a one-*wen* increase per *mu* (1 *mu* is approximately 667 square meters) to grain and land taxes, further establishing the name "One-*Wen* Garden."

In 1784, during the Qianlong period, Wang Xiyi, the magistrate of Qingpu, enriched the garden with new features, including Yongcui Pavilion, Xiyu Bridge, Zhuojin Jetty, Yinxi Pavilion, the Flower God Shrine, Binghu Pond, Yueguo Gallery, Canxia Pavilion, and Xiele Tower. At this time, the garden was officially named "Lingyuan Garden."

In 1798, during the Jiaqing period (1796–1820) of the Qing dynasty, Liu Yunfang, the Education Commissioner of Jiangsu, visited Qingpu at the invitation of Magistrate Yang Dongping. While enjoying a banquet in the garden, Liu noticed a stream flowing through it, with two ponds encircling a rockery. This

喜雨桥、米拜亭、冰壶、清虚景泰、天光云影、迎仙阁、虬龙洞、白云坞、清籁山房，昔日的园中盛景都恢复了原貌，成为青浦城内适宜休闲游憩之地。

1911年，辛亥革命，青浦光复，城隍庙不再祀典，灵园改为公园，曲水园从此成为民众赏景、游览、休闲、聚会的好去处。1926年3月，为纪念孙中山先生，曲水园更名为"中山公园"，凝和堂改建为中山纪念堂。

1949年后，曲水园沿用"中山公园"之名，1980年恢复"曲水园"旧名。1984—1986年，上海市人民政府拨款93万元进行全园整修，同济大学陈从周教授为舟居非水、迎仙阁、夕阳红半楼、有觉堂、御书楼、得月轩、坡仙阁等景点题匾。其后的20余年中，曲水园几经修缮，其中尤以2005—2007年为最，当时的市、区两级政府共拨款1500万元，再次进行了全园整修。通过修缮建筑、优化布局、增添新景、调整绿化、完善功能、提升品位，历史上的24景中，除白云坞、米拜亭、清虚静泰外，都得到了保护和修复，并在历史传承的基础上有所发展，格局渐趋完善。

曲水园示意图（上海曲水园提供）
Schematic Diagram of Qushui Garden
(Provided by Shanghai Qushui Garden)

使刘云房来到青浦，应知县杨东屏之邀，在园中宴饮。刘见园内一溪贯之、两池绕山，便取王羲之《兰亭集序》中"流觞曲水"之意，更园名为"曲水园"，沿用至今。

清咸丰十年（1860），太平天国军队占领青浦。清同治元年（1862），清军与洋枪队联合攻城，曲水园毁于炮火。清光绪十年（1884）开始重建，先建舟居非水、夕阳红半楼、孝子堂。光绪十三年（1887）建御书楼。光绪十五年（1889）建有觉堂、得月轩。光绪十六年（1890）筑围墙、建正门。光绪十八年（1892）建坡仙阁（初建时为亭）、玉字亭、涌翠亭、喜雨桥，

曲水园荷花池旧照（上海曲水园提供）
Historical view of Lotus Pond in Qushui Garden (Provided by Shanghai Qushui Garden)

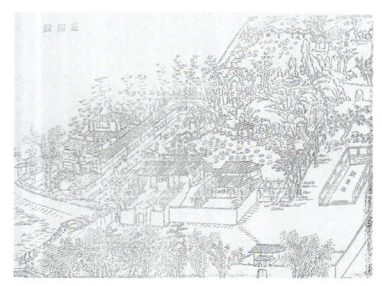

曲水园图（《青浦县志》灵园图，上海曲水园提供）
Qushui Garden Map (*Gazetteer of Qingpu County*, Lingyuan Map, provided by Shanghai Qushui Garden)

并购园东地凿放生池。光绪二十年（1894）建恍对飞来厅。光绪二十三年（1897）建凝和堂。光绪二十四年（1898）建花神祠、冰壶。清宣统二年（1910）建镜心庐。

至此，园中新旧共凑24景，分别是：凝和堂、夕阳红半楼、环碧楼、迎曦亭、玉字廊、二桥、有觉堂、舟居非水、得月轩、镜心庐、坡仙阁、濯锦矶、花神堂、恍对飞来、小濠梁、

177

历史沿革

Garden History

在漫长的岁月变迁中,曲水园屡遭兵火,数遇动乱,几度更名,历经多次损毁重建,但始终没有停下东扩和北延的脚步。特别是1949年后的两次大规模整修,使曲水园旧貌换新颜。

曲水园初建于清乾隆十年(1745),曾是青浦城隍庙的附属园林。跨清乾隆、嘉庆、道光、咸丰、同治、光绪、宣统七朝(1736—1911),越民国时期,经新中国诞生,历改革开放,见证着时代的沧桑变幻。

青浦城隍庙初建于明万历年间(1573—1620),香火甚旺。相传清雍正十二年(1734)的春天,有位远方云游的得道高僧来到青浦,观天象,知雍正王朝气数将尽,新皇将立,即与城隍庙主持道长互通,以城隍庙善男信女的捐款及高僧云游所积攒的钱款共建城隍庙"灵园"。于是,在清乾隆元年(1736)开始选址、规划、做蓝图,以宗教理念为基础,用9年时间定下灵园的整体规划,于乾隆十年(1745)建有觉堂、得月轩等,乾隆三十二年(1767)建凝和堂、花神堂并挖河道,乾隆四十九年(1784)凿池垒山、植莲种树、筑亭建桥,至乾隆末年气数耗尽,灵园中的三角亭以茅草结顶,草草完工。是时,年号更迭,高僧圆寂。虽然传说中的细节真伪已很难细考,但是有据可查的是,青浦城隍庙向来有庙捐,谓之"一文愿",用以修庙。当时园、庙一体,修园经费亦来自"一文愿",故百姓便称此园为"一文园"。坊间还有一种说法,曲水园建成约40年后,青浦士绅联名呈请,获准在粮税、地丁税项下每亩附加制钱一文,集款扩建园林,故有"一文园"之称。

乾隆四十九年(1784),王希伊任青浦知县,建涌翠亭、喜雨桥、濯锦矶、寅曦亭、花神祠、冰壶、鱼乐国轩、餐霞亭、偕乐楼等,该园正式被命名为"灵园"。

清嘉庆三年(1798),江苏学

曲水园
Qushui Garden

　　曲水园位于青浦老城厢东北隅公园路北侧，东临护城河，西接城隍庙，占地约 30 亩，有着 270 多年的历史。曲水园以水见长，以小巧玲珑、典雅古朴著称，它与嘉定"秋霞圃"、南翔"古猗园"、黄浦"豫园"、松江"醉白池"合称为"上海五大古典园林"。2014 年 4 月，曲水园被公布为上海市文物保护单位，是首批上海市"文明公园"。2022 年 8 月 1 日起，该园免费向公众开放。

Qushui Garden, located in the northeast corner of Qingpu's old city, sits along the north side of Gongyuan Road, with the moat to its east and the City God Temple to its west. Covering approximately 30 *mu* (20 000 square meters), the garden has a history spanning more than 270 years. Famous for its graceful waterscape, Qushui Garden captivates visitors with its refined, timeless charm. Alongside Jiading's Qiuxia Garden, Nanxiang's Guyi Garden, Huangpu's Yu Garden, and Songjiang's Zuibaichi Garden, it ranks among Shanghai's top five classical gardens. In April 2014, it was announced as Shanghai city-level protected site and recognized as one of the city's first "Model Parks." Since August 1, 2022, Qushui Garden has offered free entry to the public.

清籁山房
Qinglai Studio

怡景览胜，今朝传承——曲水园

Relishing Scenic Beauty in Modern Times: Qushui Garden

年代：清代
地址：上海市青浦区公园路 612 号
占地面积：约 20 000 平方米
保护级别：上海市文物保护单位

Era: Qing dynasty
Address: No. 612 Gongyuan Road, Qingpu District, Shanghai
Area: approximately 20 000 square meters
Protection level: Shanghai city-level protected site

曲水园鸟瞰图
Aerial view of Qushui Garden

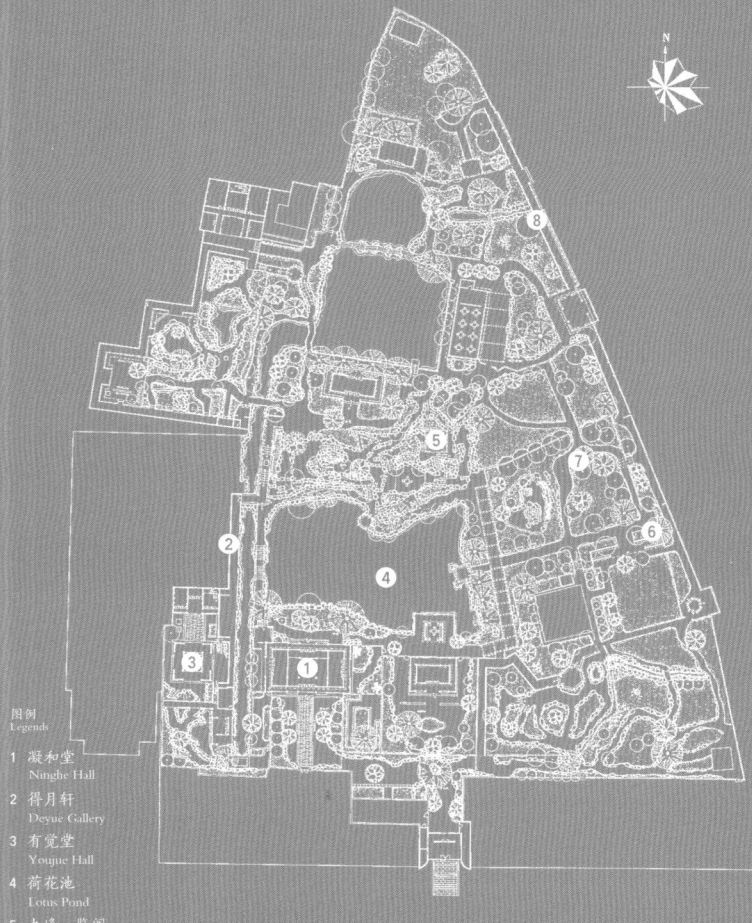

图例
Legends

1 凝和堂 Ninghe Hall
2 得月轩 Deyue Gallery
3 有觉堂 Youjue Hall
4 荷花池 Lotus Pond
5 九峰一览阁 Jiufeng Yilan Tower
6 听橹阁 Tinglu Belvedere
7 机云亭 Jiyun Pavilion
8 绿波廊 Lübo Corridor

曲水园

Qushui Garden

Zuibaichi Garden's transformation from a private garden into a public space began when it was acquired by the orphanage during the Jiaqing period (1796–1820) and partially repurposed as a Rent Collection Office. Thanks to the efforts of the orphanage leaders, the garden survived periods of unrest, including the Taiping Heavenly Kingdom Movement, and received ongoing repairs funded by patrons. Visitors today can still see six weathered stone tablets in Jianshan Hall inscribed with "Under the Guardianship of the Orphanage." Zuibaichi Garden also holds an important place in modern Chinese history. On December 26, 1912, Dr. Sun Yat-sen, the founding father of modern China, visited Songjiang and attended a luncheon at Xuehai Hall, hosted by local officials and community leaders. During his visit, Sun gave a significant speech commemorated by couplets on the hall's pillars: "The echoes of President Sun's revolutionary speech will endure through the ages" and "The gathering of Songjiang's elite at Xuehai Hall is an honor celebrated for millennia." On June 19, 1923, Songjiang's first Communist Party member, Hou Shaoqiu, together with Zhu Jixun and other local patriots, established the "Songjiang National Salvation Association" at Chishang Cottage in Zuibaichi Garden. The association aimed to rally the community around four goals: to overthrow warlords, end imperialist influence, eliminate bureaucratic corruption, and promote social welfare — initiatives that marked a turning point in Songjiang's revolutionary history. After the outbreak of the War of Resistance against Japanese Aggression in 1937, Zuibaichi Garden suffered extensive damage. The forced changes to its architectural style reflect this painful chapter of history. Songjiang's rich cultural legacy is deeply embedded in the garden's landscape, making Zuibaichi a silent witness to the many chapters of its past.

Today, Zuibaichi Garden stands as a cultural treasure of Jiangnan, cherished not only in Songjiang but across Shanghai. Visitors can stroll around its serene pond, explore exquisite architecture, and experience the charm of a classic Jiangnan water town. The garden offers a window into history, sharing stories of prominent figures and pivotal moments that have shaped modern Shanghai. As a preserved chronicle of Songjiang's centuries-old heritage, Zuibaichi Garden continues to thrive through thoughtful restoration, blending history with vitality. Its rich legacy promises to inspire future generations, harmonizing with Shanghai's growth and enriching its cultural landscape.

辍耕亭
Chuogeng Pavilion

听橹樓
Tinghe Gazebo

雪海堂内景
Interior of Xuehai Hall

上海近代的风云变幻。它既是一部凝固的史书,见证了松江数百年来的人文积淀,也是一段活态的乐章,在修缮改造中不断焕发新的生命。未来,它将继续以自身之文化底蕴为松江乃至上海的发展注入源源不断的活力,与时代脉搏共振,与城市建设共生。

> **参观指南 Visiting Guide**
>
> 开放时间:06:00—17:00(16:30 停止入园)。
> 交通信息:地铁 9 号线至醉白池站 3 号出口;公交莲松专线(莲花路地铁站)—松江招商市场站转 9 路、12 路、19 路至醉白池站。
>
> Opening hours: 06:00-17:00 (16:30 Last admission).
> Traffic information: Metro Line 9 to Zuibaichi Park Station (Exit 3); Liansong Special Line Bus (Lianhua Road Metro Station) - Songjiang Zhaoshang Market Station, transfer to Bus 9, 12, 19 to Zuibaichi Park Station.

史海钩沉

Historical Stories

自清嘉庆年间（1796—1820）被育婴堂购得并辟为征租所后，醉白池从私家园林逐渐变为公共事务空间，在历任育婴堂主事的庇护下，醉白池在太平天国动乱中幸免于难，也因他们的出资得到修葺与爱护。如今，人们在见山厅内还可看到6块刻有"育婴堂奉文保管界"且字迹斑驳的石碑。此外，醉白池还见证了松江地区诸多政治事件。1912年12月26日，孙中山先生视察松江，于12月27日下午在雪海堂出席县教育会、城市公所、县议事会和松江商会等四个团体共设的欢迎午宴，并作重要演讲。雪海堂立柱上的堂联，"孙文总统演讲革命余音绕梁永世，雪海一堂会见松人殊荣传誉千秋"即是对此事件的纪念。1923年6月19日，松江的第一名共产党人侯绍裘与朱季恂等人联合各界爱国人士，在池上草堂成立"松江救国同志会"，"以联合本地各界以民治精神达民族自决为救国目的"，公开提出打倒军阀、打倒国际帝国主义、铲除官僚政治、提倡社会服务四项信条，掀起了松江革命的新篇章。1937年抗战爆发后，醉白池遭到严重破坏，建筑风格的强制改变展现了这段血泪交加的历史。可以说，松江的人文历史深深嵌置在醉白池的园林景观中，使醉白池成为一段段历史的无声见证者。

今日之醉白池已成为松江乃至上海展示江南文化的一张重要名片，在此可赏一泓清池，访雕梁画栋，品悟江南水乡的人文风韵，也可听历史故事，沐伟人之光，体味

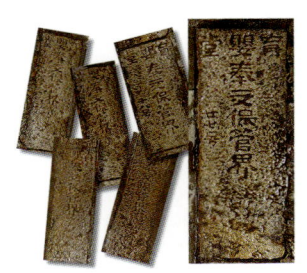

"育婴堂奉文保管界"碑（上海醉白池公园提供）
Stone tablets inscribed with "Under the Guardianship of the Orphanage" (Provided by Shanghai Zuibaichi Garden)

疑舫（上海醉白池公园提供）
Yi Stone Boat (Provided by Shanghai Zuibaichi Garden)

Zuibaichi Garden, one of the Shanghai's five classical gardens, is celebrated for its well-preserved historical character, unique architecture, and tranquil water features. Established during the Qing dynasty, much of the garden's original layout dating back to the Kangxi (1662–1722) and Yongzheng (1723–1735) periods remains intact, offering visitors a glimpse into traditional Jiangnan garden design.

At the heart of the garden lies a long, square pond, known for its natural simplicity — without islands or bridges — to create an open, flowing feel. To the north, the Four-Sided Hall is surrounded by water, with nearby Chishang Cottage adding to the harmony of the scene. Although the Old Tree Pavilion no longer stands, a corridor along the water's edge still provides a timeless path that connects visitors to the past.

The garden's inspiration comes from Tang poet Bai Juyi's *Pondside*, which envisions a serene retreat with "a ten-*mu* estate and a five-*mu* garden, a pond of water, a thousand bamboo stalks … a hall, a courtyard, a bridge, and a boat … cranes, rocks, water chestnuts, and lotus blossoms … " This peaceful setting echoes in Zuibaichi's graceful layout, which combines ponds, rustic bridges, stone boats, lotus flowers, and quiet rock formations to create a natural, village-like charm.

Despite some modern updates, including structures like Xuehai Hall and Magnolia Courtyard, Zuibaichi Garden preserves the essence of a traditional Jiangnan village. Visitors can enjoy a sense of timeless beauty, where one might feel, "in this tranquil setting, I could spend my final days."

价值特色

Features and Significance

醉白池有建筑可览，有花木可品，有池水可赏，但其最大的价值特色在于其较为完整地延续了历史风貌和意蕴，这在上海五大古典园林中是可圈可点的。在清康熙（1662—1722）、雍正（1723—1735）年间，醉白池"池方而长，可三四亩，据宅之右。池东有老榆槎枒，二百年物，轩荫其下，临流而坐，曰老树轩，贯以长廊。池西畎亩连亘，限之以篱，篱疏可眺也。池南两三人家，窗户映带，妇孺浣汲，望若画图。池北，堂临之，敞其四面。堂北与西，竹石列环。又北，则池尾绕而东。又北有隙地，可构屋，莳花木，亦篱限其外焉……"，已与今日相去不远。这种风貌格局可与醉白池的灵感来源——白居易《池上篇》中"十亩之宅，五亩之园。有水一池，有竹千竿……有堂有庭，有桥有船……灵鹤怪石，紫菱白莲……妻孥熙熙，鸡犬闲闲……"的描述相对照，呈现了天然、质朴的田园风光。

200多年后，醉白池内的建筑有所重修、改建，水系格局也有变化，但内园的景观格局仍带有清初至清中叶的风貌特征。例如，如今的内园仍旧以"方而长"的池为中心，水中不设岛屿、桥梁为分隔，保持着自然意蕴；池北有四面厅，水体绕四面厅而东，只是增加了池上草堂；堂北依旧构屋莳花木，只是不再用篱笆进行隔断，而是划定边界，以院墙进行围合。池东"老树轩"虽不存，但长廊依旧；池西与池南变化较大，原先以篱笆这种轻隔断制造出的开敞田园风光，为"雪海堂""玉兰院"及"仓房"所取代，但即使以后期建筑的填建制造了空间的闭合，醉白池的核心"一池""一堂"依旧存在，池边"桥""舫""怪石""莲"景观仍然保留，并且营造着能让人"优哉游哉，吾将终老乎其间"的环境氛围，赋予了醉白池"小桥流水人家"的村野之趣。

除了建筑景观,醉白池内还有历代碑刻值得一看。其中最著名的当为《云间邦彦画像》刻石,共30块,嵌于池南长廊壁上,共刻有明代松江府91位乡贤,包括著名书画家董其昌、陈继儒,状元钱福,宰相徐阶,礼部尚书陆树声,刑部尚书张鏊,工部尚书潘恩,松江诗派主将陈子龙、夏允彝等人,原嵌于府学明伦堂壁上,后于1941年移置醉白池。

在宝成楼南侧有赵孟頫所书苏轼《前后赤壁赋》碑文,原40余方,行书大字,得李北海体势,现仅存22方。此外,还有吴道子观音画像石刻、董其昌"韩范先声"石刻、方孝孺"正心诚意"碑、《醉白池记》碑等。这些碑刻大多从他处移来,既丰富了醉白池的文化底蕴,也展现了松江的历史面貌。

Beyond its architectural beauty, Zuibaichi Garden is home to a collection of notable steles. The most famous set, embedded in the corridor wall to the south of the pond, consists of 30 slabs depicting 91 prominent Ming dynasty figures from Songjiang. Among them are celebrated artists like Dong Qichang and Chen Jiru, Principal Graduate Qian Fu, Prime Minister Xu Jie, and esteemed ministers such as Lu Shusheng, Zhang Ying, and Pan En. The series also honors leaders of the Songjiang Poetry School, Chen Zilong and Xia Yunyi. Originally displayed on the wall of Minglun Hall in the Songjiang Confucian School, these steles were relocated to Zuibaichi Garden in 1941.

South of Baocheng Tower is a stele bearing Su Shi's famous *First and Second Odes on the Red Cliff*. Su Shi, a renowned poet, calligrapher, and statesman of the Song dynasty, is celebrated for his literary and artistic achievements. The calligraphy on the stele is by Zhao Mengfu, a distinguished Yuan dynasty calligrapher, painter, and scholar. Originally, this collection consisted of over 40 large-character steles in the style of Tang dynasty calligrapher Li Yong; today, 22 of these remain. Other notable pieces include an engraving of the Bodhisattva by the renowned Tang dynasty painter Wu Daozi, as well as works by Dong Qichang and Fang Xiaoru, a distinguished scholar of the early Ming dynasty. Many of these steles were moved from various locations, enhancing the cultural richness of Zuibaichi Garden and showcasing Songjiang's deep historical legacy.

碑刻艺术
Collection of Notable Steles

上：池南廊壁上的《云间邦彦画像》(上海醉白池公园提供)
U: Prominent Ming dynasty Figures from Songjiang, embedded in the corridor wall to the south of the pond (Provided by Shanghai Zuibaichi Garden)
中：《云间邦彦画像》(部分)
M: Prominent Ming dynasty Figures from Songjiang (Detail)
下：《前赤壁赋》(部分)
D: First Ode on the Red Cliff (Detail)

继续向东,可见见山厅和宝成楼。见山厅原是供主人停放轿子的地方,今为醉白池首任园主顾大申生平展示厅,宝成楼则是一座面阔五间的二层雕花楼,为原主人住宅。再向东为醉白池的原大门。大门是三开间门厅,鱼龙脊,阑额上雕刻精美,有人物、建筑、树木、花卉、假山等。图中人物神态灵动,或读书弈棋,或品茗赏花,再现了明清士大夫的生活和情趣。

Continuing east, visitors encounter Jianshan Hall and Baocheng Tower. Jianshan Hall, named after the garden's first owner, Gu Dashen, who went by the literary name "Jianshan," was originally a resting place for the owner and his sedan chair. Today, it houses an exhibition on Gu Dashen's life and his role in the garden's history. Adjacent to Jianshan Hall is Baocheng Tower, a two-story building with intricate carvings and a five-bay facade. It originally served as the owner's residence. Further east, visitors reach the original main gate of Zuibaichi Garden, a three-bay entrance hall featuring a roof ridge decorated with fish and dragon motifs. The architrave is adorned with finely carved scenes of figures, buildings, plants, and rockeries. These lively carvings depict scholars reading, playing chess, drinking tea, and admiring flowers — reflecting the cultured lifestyle and leisurely pursuits of Ming and Qing scholars.

宝成楼
Baocheng Tower

莲叶东南榭（大湖亭，上海醉白池公园提供）
Lianye Dongnan Gazebo (Great Lakeside Pavilion, provided by Shanghai Zuibaichi Garden)

半山半水半书窗亭
Half-Mountain, Half-Water, Half-Window Pavilion

醉白池池东有廊，有花露涵香榭和莲叶东南榭穿插其中。花露涵香榭，又称"小湖亭"，系清嘉庆年间（1796—1820）所造，因该亭西北边植有百年紫牡丹得名。明代著名书画家董其昌曾亲手栽种牡丹，赠予奉贤邬桥好友金学文，以庆其新屋落成之喜，嘉庆年间，醉白池园主为纪念董其昌，将金家的牡丹分株种植于此。每年清明时节牡丹盛开时，人们可坐于亭中闻香赏景。莲叶东南榭又称"大湖亭"，亦造于嘉庆年间，因夏日莲叶似波，涌入东南亭中而得名。在此观荷赏鱼，最为适宜。池南有环廊。西南角有一处名为"半山半水半书窗"的六角湖亭，建于清光绪二十五年（1899）。因其一半倚于池岸、一半悬于池上、亭的东部设窗而得名。

East of Zuibai Pond stretches a corridor with two gazebos along its length: Hualu Hanxiang Gazebo and Lianye Dongnan Gazebo. Built during the Jiaqing period (1796–1820), Hualu Hanxiang Gazebo, or the Small Lakeside Pavilion, is named for the purple peonies planted nearby. These peonies honor a tradition dating back to Dong Qichang, the famous Ming dynasty artist, who gifted peonies to his friend Jin Xuewen in celebration of Jin's new home. To commemorate Dong's legacy, peony cuttings from Jin's garden were planted here during the Jiaqing period. Around the Qingming Festival, typically in April, visitors can sit in the gazebo, enjoying the sight and fragrance of the blooming peonies. Lianye Dongnan Gazebo, also known as the Great Lakeside Pavilion, was also built during the Jiaqing period. Its name reflects the summer scene as lotus leaves float toward the southeast pavilion. This is an ideal spot for viewing lotus blooms and fish in the pond. South of Zuibai Pond, a winding corridor leads to the hexagonal pavilion known as "Half-Mountain, Half-Water, Half-Window." Built in 1899 during the Guangxu period, this pavilion is uniquely designed with one half resting on the pond bank and the other suspended over the water, offering lovely views through windows on its eastern side.

花露涵香榭（小湖亭）
Hualu Hanxiang Gazebo (Small Lakeside Pavilion)

百年牡丹（上海醉白池公园提供）
Century-old peonies (Provided by Shanghai Zuibaichi Garden)

柱颊山房往北有建于"谷阳园"文澜堂遗址上的乐天轩，为1927年新造。东北处有疑舫，为清光绪二十三年（1897）重修，虽名"舫"，但更似一"轩"，以山面为入口，北面伸入池中，似池中之舟，蕴含传统文人对不系之舟的精神追求。

North of Zhujia Studio is Letian Gallery, built in 1927 on the site of Wenlan Hall from the original Guyang Garden. To the northeast is Yi Stone Boat, rebuilt in 1897. Although it resembles a gallery rather than an actual boat, it extends gracefully into the pond. The name reflects the traditional scholar's ideal of a "free-floating boat," symbolizing a life detached from worldly concerns.

乐天轩
Letian Gallery

醉白池一隅
A corner of Zuibai Pond

池西可见卧树轩，建于 1926 年，此处有一株古女贞树卧于轩旁，至今已有百余年。池北为池上草堂和柱颊山房。池上草堂是内园的中心建筑，建于清宣统元年（1909），整座建筑凌空于池上，四面隔扇，周有轩廊，古朴雅致，曲栏横槛；堂上原来悬挂清四王之一——太仓王时敏所题"醉白池"匾额，"文革"时期遭毁，如今的匾额由国画大师程十发先生补题于 1972 年，风格近唐隶。柱颊山房为明式建筑风格，因四面皆为厅窗，俗名"四面厅"，此处曾为云间书画派鼻祖、南京礼部尚书董其昌吟诗作画、以文会友之地。

To the west, the Woshu Gallery, built in 1926, sits beside a century-old glossy privet tree. On the north side of the pond are Chishang Cottage and Zhujia Studio. Chishang Cottage, built in 1909, is the focal point of the Inner Garden, positioned over the pond with partition doors on all sides and an open corridor encircling it. Originally, the cottage displayed a plaque with the words "Zuibai Chi" (Drunken Bai Pond) inscribed by Wang Shimin, one of the Qing dynasty's renowned "Four Wangs." Although the original plaque was destroyed during the Cultural Revolution, it was replaced in 1972 by artist Cheng Shifa, who inscribed it in a classic Tang-style clerical script. Zhujia Studio, known as the "Four-Sided Hall" for its windows on all sides, was a favorite retreat of Dong Qichang, the renowned Ming dynasty calligrapher and painter who founded the Yunjian School. Dong often came here to write poetry and paint, leaving a lasting influence on the garden's artistic heritage.

池上草堂侧影（上海醉白池公园提供）
Side view of Chishang Cottage (Provided by Shanghai Zuibaichi Garden)

柱颊山房内景
Interior of Zhujia Studio

内园
Inner Garden

雪海堂外景
Exterior of Xuehai Hall

　　内园如今以雪海堂为序。雪海堂为五开间建筑，堂前有庭院，后于 20 世纪 70 年代开挖围砌芳池，如今池内栽植睡莲名品。继续往里走，可见内园的核心景观——醉白池。醉白池平面近方形，周圈以黄石堆砌，池中遍植荷莲，环池一圈建有亭台阁榭。

The Inner Garden centers around Xuehai Hall, a five-bay structure with a front courtyard. In the 1970s, a pond with blooming water lilies was added, enhancing the garden's tranquil atmosphere. Deeper inside lies the garden's heart — Zuibai Pond. Nearly square and bordered by yellow stones, the pond is filled with lotus plants and surrounded by waterside structures.

走进外园,迎面可见一面由苏州陆墓一带收集而来的清代方砖制成的巨型砖雕照壁,展现的是清代醉白池及松江古城风貌。入园后,左侧可见雕花厅。厅坐北朝南,为明成化年间(1465—1487)南安知府张弼后裔张祖南豪宅的一部分,建成于清代晚期。现存二厅二厢,前后厅均为五开间,东西厢房均为三开间,彼此间有廊相通。梁枋门窗雕花卉与《三国演义》故事,为松江现存古建筑所罕见。主干道右侧可见五色泉和泼水观音,还可散见一些新造的亭台水榭。

Entering into the Outer Garden, visitors encounter a large brick screen wall made of Qing dynasty bricks collected from Suzhou, featuring carvings that depict Zuibaichi Garden and ancient Songjiang as they appeared in the Qing dynasty. To the left within the garden, Diaohua Hall now faces south and features two main halls and two side wings, all connected by corridors. Once part of the residence of Zhang Zunan, a descendant of Ming dynasty Prefect Zhang Bi, the hall features beams, brackets, and windows intricately carved with floral motifs and scenes from *Romance of the Three Kingdoms*, a classic Chinese novel. This level of detailed carving is rare among Songjiang's historical buildings, offering visitors a glimpse into traditional Chinese craftsmanship and literary heritage. On the right side of the main path, visitors will come across the Wuse Spring and the fascinating "Water-Revealed Guanyin," a stone where the image of the Bodhisattva Guanyin gradually appears when splashed with water. Nearby are beautifully designed waterside structures that invite peaceful reflection.

五色泉
Wuse Spring

外园
Outer Garden

上：西大门入口照壁
U: Screen wall at the west main entrance
下：雕花厅
D: Diaohua Hall

园景撷趣

Garden Scenery

盆景园
Bonsai Garden

　　如今，游客主要从公园西大门进入外园，行至最深处可达内园。外园的主要景观有雕花厅、五色泉和泼水观音等，整体风格简约清新。内园保存有以醉白池为中心建造的亭台廊榭，如雪海堂、池上草堂、柱颊山房等，整体风格古朴清幽，是"春访牡丹夏观荷，秋来赏菊冬瞻梅"的好去处。

Visitors typically enter the Outer Garden through the west entrance before continuing into the Inner Garden. The Outer Garden, featuring Diaohua Hall, Wuse Spring, and the Water-Revealed Guanyin, has a simple yet refreshing charm. At the heart of the Inner Garden is Zuibaichi Garden, surrounded by pavilions, halls, and corridors such as Xuehai Hall, Chishang Cottage, and Zhujia Studio. With its serene beauty, it offers the perfect setting to enjoy peonies in spring, lotus blossoms in summer, chrysanthemums in autumn, and plum blossoms in winter.

艺术碑廊
Stele Gallery

Courtyard" to the south of Xuehai Hall and "Shanglu Garden" in the southeastern section. In 1983, an old warehouse was transformed into a stele gallery, displaying steles collected from across Songjiang. The intricately detailed Diaohua Hall, originally part of the late Qing Zhang family residence on Xita Lane in Songjiang, was also relocated to the Outer Garden. Meanwhile, the Rent Collection Hall and Baocheng Tower in the Inner Garden were rebuilt, and the garden's main gate was restored, shaping Zuibaichi Garden as visitors see it today.

was well-regarded as a poet, artist, and water management expert. After facing challenges in his official career, he returned to his hometown and devoted himself to cultivating the garden, bringing his artistic and literary sensibilities to its design. Inspired by the Tang dynasty poet Bai Juyi, who was celebrated for his love of nature and wine, Gu Dashen redesigned the garden and named it "Zui Bai" (Drunken Bai) in his honor. By the pond, he constructed Zuibaichi Garden as a gathering place, where he hosted literary gatherings with prominent local scholars and artists. After Gu's passing, the garden changed hands several times. In the early Qianlong period of the Qing dynasty, it was acquired and restored by Gu Sizhao, a former Assistant Instructor at the Confucian School in Dantu County. Gu Sizhao continued the garden's tradition of hosting literary gatherings, inviting fellow scholars to enjoy the serene surroundings and compose poetry, maintaining the garden as a vibrant cultural hub.

After Gu Sizhao's passing, the garden gradually declined. In 1797, during the Jiaqing period, the Songjiang orphanage acquired the property and converted part of it into a rent collection office, while some buildings became warehouses and offices, including the newly built Rent Collection Hall (now called "Jianshan Hall"). Other areas served as meeting spaces for philanthropists and officials supporting the orphanage, who funded repairs to the garden's structures, such as Baocheng Tower, the Great Lakeside Pavilion, the Small Lakeside Pavilion, and the long corridor. Further restorations were undertaken during the Guangxu period (1875–1908), built storage rooms added south of the pond, and a hexagonal pavilion constructed to the southwest. In 1909, new structures like Chishang Cottage and Xuehai Hall were added along the water's edge, while additional plantings around the pond enhanced the garden with seasonal flowers and fragrant blooms, offering visitors a sensory experience.

During the Republic of China period, Zuibaichi Garden continued to develop, with the addition of the Letian Gallery on the north side of the pond in 1927. However, in 1937, after the outbreak of China's resistance war against Japanese aggression, Songjiang was occupied, and Zuibaichi Garden was taken over by Japanese forces. Most of the garden's structures were converted into Japanese-style buildings, except for Xuehai Hall and Baocheng Tower. Many plaques, couplets, trees, and flowers were destroyed causing significant damage. After the war, ownership of the garden returned to the orphanage, though various organizations, including the Seven Counties Relief Association and the Kuomintang Nationalist Traffic Police, occupied the buildings until Songjiang's liberation. In 1958, the garden was transferred to the local government, which expanded it by acquiring an additional 60 *mu* (40 000 square meters) of land to the west, creating what is now known as the Outer Garden. The original over 16 *mu* (approximately 11 100 square meters) area is now referred to as the Inner Garden.

In 1972, Zuibaichi Garden underwent minor repairs, and in 1981, Shanghai Municipal People's Government funded a full restoration of its historical buildings, including the halls, pavilions and stone boats, restoring their original Ming and Qing dynasty style. Two new scenic areas were added: "Magnolia

Zuibaichi Garden was built on the site of the ancient Guyang Garden, originally established during the Northern Song dynasty by Zhu Zhichun, who took the name "Master Guyang." The name Guyang, meaning "south of the Gushui River," is inspired by a line from the poetry of Lu Ji, a renowned literary figure from Zhu's hometown of Songjiang. Lu Ji and his brother Lu Yun were celebrated patriots and literary icons of the West Jin dynasty, whom Zhu deeply admired. In his own poem, *Lake Retreat*, Zhu reflects the garden's simplicity and beauty with lines like, "Ten acres of open lake water, my humble thatched hut tucked away … here, I forget the world, and my heart has no desire to leave."

During the Ming dynasty, the garden featured a tranquil pond shaded by an ancient elm, with a pavilion called the "Old Tree Gallery" nestled beneath its branches. To the north stood Shenliu Dushu Hall, alongside other structures that completed the peaceful setting. In the early Qing dynasty, during the Shunzhi (1644–1661) and Kangxi (1662–1735) periods, the garden was acquired by Gu Dashen, a native of Huating County in Songjiang. Gu, who attained the *jinshi* title after passing the 1652 imperial palace examination,

玉兰院
Magnolia Courtyard

民国时期，醉白池仍有修筑。今池北"乐天轩"即于 1927 年所建。1937 年抗战爆发，11 月松江沦陷，醉白池被日军占据，将除雪海堂、宝成楼之外的堂轩亭榭均改成日式建筑，焚毁匾额楹联，摧残树林花卉，对醉白池造成较大破坏。抗日战争胜利后，醉白池产权归育婴堂所有，但园内建筑被七县救济会、国民党交通警察部队等占用，直至松江解放，被部队接管。1958 年，醉白池移交地方政府，征购醉白池西土地 60 亩（40 000 平方米），扩建为公园，也就是今天所称的外园部分。原先的醉白池部分占地 16 余亩（约 11 100 平方米），称"内园"。

1972 年，醉白池曾稍作修缮，1981 年，上海市人民政府拨款将雪海堂、宝成楼、征租厅外的其他古建筑进行全面落架翻建，将园内堂、轩、

原大门旧貌（摄于 20 世纪 80 年代，上海醉白池公园提供）
Original main gate in its earlier form, taken in the 1980s
(Provided by Shanghai Zuibaichi Garden)

亭、舫等主体建筑恢复明清风貌。同时在雪海堂南侧辟"玉兰院"，在公园东南部辟"赏鹿园"两组新园景。1983 年，改旧仓库为碑廊，用以陈列松江各处搜集的碑刻，后将松江镇西塔弄底张宅极尽工整华美的雕花厅大厅、两厢一组建筑移置外园，同时翻建内园的征租厅（今"见山厅"）、宝成楼，并修复醉白池原大门，成为我们今日所见的醉白池。

见山厅
Jianshan Hall

顾思照去世后,园子渐显颓势,清嘉庆二年(1797),松江府育婴堂购得产权,辟为"征租所",将园内部分房屋改建为仓房、仓厅和账房,建"征租厅"(今"见山厅")和临街平房作办事之用。园内其余建筑则用于支持善堂的士绅和官僚聚会议事。他们出资整理园景,并陆续重修宝成楼、大湖亭、小湖亭、长廊等。清光绪年间(1875—1908),建疑舫,在池南建仓房,在池西南建六角湖亭等建筑。清宣统元年(1909),筑水阁(池上草堂),建雪海堂,并在池畔建筑之间遍植花木,使醉白池四季皆有花可赏,有香可闻。

历史沿革
Garden History

该园建于北宋朱之纯所建的"谷阳园"遗址之上。朱氏自号谷阳先生,因仰慕松江名人陆机、陆云,遂以陆机诗句中的"仿佛谷水阳"命名该园。朱氏曾作《湖斋》一诗云:"平湖十顷水汪洋,得意茅斋且屈藏……居此倏然忘世味,此心尤懒去龙阳",从中可见当时一湖一斋的朴素意味。

明代,这里有寒塘一片,塘东老榆一株,树下为"老树轩",轩北有"深柳读书堂"等建筑。清顺治(1644—1661)、康熙(1662—1735)年间,该园由松江府华亭县人顾大申购得。顾大申名庸,字震雉,号见山,顺治九年(1652)进士,是著名的学者、诗人和水利专家,擅诗书画,为松江画派后期的主要人物之一。由于一度仕途失意,他归隐乡里,效仿魏国公韩琦仰慕白居易而筑"醉白堂"于私第之池上的举动,以"醉白"命名修建后的池上园林,与当地名流在此唱酬游宴。顾大申去世后,园子数易其主,乾隆初年由曾任丹徒县训导的顾思照购入并加以修葺,还召集了一大批文人在此唱和抒怀。

清朝末年的醉白池(上海醉白池公园提供)
Zuibaichi Garden in the late Qing dynasty
(Provided by Shanghai Zuibaichi Garden)

池上草堂(摄于20世纪50年代,上海醉白池公园提供)
Chishang Cottage, taken in the 1950s
(Provided by Shanghai Zuibaichi Garden)

池上草堂
Chishang Cottage

醉白池位于松江区人民南路 64 号，是上海五大古典园林之一，凭借着一泓清池和一腔诗意吸引着世界各地的游客到访。

Located at No. 64 South Renmin Road in Songjiang District, Zuibaichi Garden is one of the Shanghai's five famous classical gardens. Known for its clear waters and serene atmosphere, the garden draws visitors from all over the world.

醉白池
Zuibai Pond

古园诗韵，文脉流芳——醉白池
Poetic Elegance and Lasting Legacy: Zuibaichi Garden

年代：清代
地址：上海市松江区人民南路 64 号
占地面积：约 51 100 平方米
保护级别：上海市文物保护单位

Era: Qing dynasty
Address: No. 64 Renmin South Road, Songjiang District, Shanghai
Area: approximately 51 100 square meters
Protection level: Shanghai city-level protected site

醉白池鸟瞰图（上海醉白池公园提供）
Aerial view of Zuibaichi Garden (Provided by Shanghai Zuibaichi Garden)

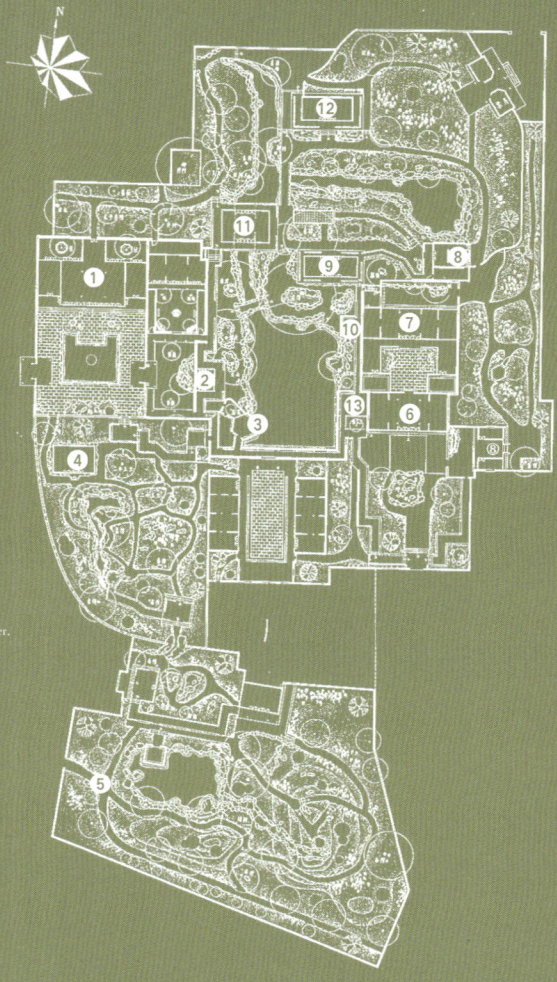

图例
Legends

1 雪海堂
  Xuehai Hall
2 卧树轩
  Woshu Gallery
3 半山半水半书窗亭
  Half-Mountain, Half-Water,
  Half-Window Pavilion
4 玉兰院
  Magnolia Courtyard
5 赏鹿园
  Shanglu Garden
6 征租厅
  Rent Collection Hall
7 宝成楼
  Baocheng Tower
8 疑舫
  Yi Stone Boat
9 四面厅
  Four-Sided Hall
10 花露涵香榭
   Hualu Hanxiang Gazebo
11 池上草堂
   Chishang Cottage
12 乐天轩
   Letian Gallery
13 莲叶东南榭
   Lianye Dongnan Gazebo

醉白池内园
Inner Garden

醉白池

Zuibaichi Garden

老正门
Old main entrance

鸢飞鱼跃轩与南厅
Yuanfei Yuyue Pavilion and South Hall

Over the past five centuries, Guyi Garden has transformed from a classical residence garden, reflecting refined scholarly tastes, into a more diverse temple garden, and ultimately into a modern public space that blends sophistication with popular appeal. Throughout its history, the garden has witnessed significant changes, endured wartime destruction, and undergone numerous restorations and reconstructions. These transformations have layered the garden with both tradition and innovation, merging classical and contemporary influences to shape its unique character.

Today, Guyi Garden thrives as a vibrant and fashionable public space, embodying the evolution of Jiangnan gardens. Its return to public life celebrates a delicate balance between historical elegance and the lively energy of popular culture.

猗园风韵

The Charm of Guyi Garden

近 500 年历史中，古猗园从彰显文人趣味的古典宅园，到多元与世俗化的城隍庙园，再到雅俗纷呈的近代园林，历经多次变迁，屡遭战火破坏，辗转之间多次修缮与重建，古猗园承载了不同时代、不同阶层、不同文化与价值的烙印，从而积淀下或传统、或变异、或古典、或现代的面貌，糅合成其独特风韵。

昨日的古典园林，今日已成为多元、时尚、开放的城市公共空间，这是江南园林流变的范式之一，我们感慨其嬗变历程，也慨叹其以雅俗纷呈的面貌回归大众。

鹤寿轩
Heshou Gallery

1 张承先. 南翔镇志 [M]. 程攸熙, 订. 朱瑞熙, 标点. 上海: 上海古籍出版社, 2003: 2.
2 另两园分别为李宜之祖父李文邦营建的三老园、叔父李流芳营建的檀园。
3 上海古猗园. 古猗园志 [M]. 上海: 上海文化出版社, 2018: 96.
4 张承先. 南翔镇志 [M]. 程攸熙, 订. 朱瑞熙, 标点. 上海: 上海古籍出版社, 2003: 58.
5 童寯. 江南园林志 [M]. 2 版. 北京: 中国建筑工业出版社, 2014: 22.

参观指南 Visiting Guide

开放时间：3—11 月，07:00—18:30（南门），05:30—19:30（北门）；12—2 月，07:00—19:00（南门），06:00—17:00（北门）。

交通信息：地铁 11 号线至南翔站，1 号口出站后向西步行 15 分钟可达，或自 2 号口出站后换乘嘉定 129 路、嘉定 111 路；公交 62 路、821 路。

Opening hours: March to November, 07:00-18:30 South Gate), 05:30-19:30 (North Gate); December to February, 07:00-19:00 (South Gate), 06:00-17:00 (North Gate).
Traffic information: Metro Line 11 to Nanxiang Station, exit at Exit 1 and walk west for 15 minutes, or exit at Exit 2 and transfer to Jiading Bus 129, 111; Bus 62, 821.

鸳鸯湖远景
Distant view of Yuanyang Lake

古猗园荷花
Guyi Garden lotus

Though time has passed and its original creators are long gone, Guyi Garden endures, continually renewing itself through the centuries. After the renewal of Guyi Garden, the Yi Garden area, rich with historical artifacts, radiates serenity and grace. The Quxi Heying area, surrounded by water on three sides, presents an idyllic, picturesque setting. The Huaxiang Xianyuan area offers expansive views and a bright, refreshing landscape. The Youhuang Yanyue area, featuring bamboo, creates a tranquil atmosphere where bamboo and stones harmoniously complement each other. Guyi Garden, originally a scholar's retreat, stands out for its distinctive features in five major aspects. First, the garden exudes a strong literary ambiance. Zhu Sansong, inspired by Bai Juyi's *Pondside* poem, envisioned "a pond of water, where a thousand bamboo rise," creating the ideal setting for scholarly contemplation. The line "a thousand bamboo, polished green and bright" symbolizes the scholar's pursuit of refinement. This literary spirit is still palpable today, with various bamboo species gracing the garden — some tall and stately, others gently swaying — harmonizing with the water, sky, and garden architecture. Second, the garden's landscape design is notable. The areas around Yiye Hall, Xiaosonggang, and Xi'e Pond preserve their original layout, offering a scene reminiscent of a classical Chinese painting. Winding lakes, layered reflections of hills and ponds, and intricate scenery create a living landscape painting. From Yiye Hall, visitors can view rippling waters that wind into secluded corners, opening up to broad vistas of pine-covered hills, scattered stones, and faint reflections of distant landscapes. Third, Guyi Garden features a rich array of artistic works,

古猗园一隅
A corner of Guyi Garden

including plaques, couplets, steles, and brick carvings. These elements not only enhance the architecture but also harmonize with the surrounding natural scenery, reflecting the garden's aesthetic elegance. Fourth, the garden's value is deepened by its layered history. Shaped over centuries of continuous development, Guyi Garden offers an evolving space that bridges past and present, enriching visitors' appreciation of its historical significance. Finally, the garden comes to life through its integration of flowers and trees. Seasonal blooms — from spring's peonies and summer's lotus to autumn's golden osmanthus and winter's wintersweet — infuse the garden with vibrant color. The plants blend seamlessly with the landscape, embodying the harmonious union of nature, artistry, and thoughtful design.

南厅侧影
South Hall silhouette

不系舟
Buxizhou (Stone Boat)

戏鹅池
Xi'e Pond

景观起到画龙点睛的作用。四在价值之叠加。保存至今的园林遗产，是历史漫长发展过程中不断叠加的成果，动态的变化过程，数百年可追溯的历史，连通现实的构成，有助于拓展人们对园林价值的整体认知。五在花木之融入，园林无花木则无生气[5]，古猗园树木植被别具画意。自春迄冬，园内四时美景皆不相同。除了建园以来遍植的各类竹子，园内春有牡丹，

董其昌题"华岩墨海"（逸野堂内）
Dong Qichang's Inscription "Hua Yan Mo Hai" (Inside the Yiye Hall)

夏有荷花，秋有金桂，冬有蜡梅。花木与园景，形与神、意与境自然融合。

价值特色
Features and Significance

历史流转，昔人已去，而园林犹在，历久弥新。更新后的古猗园，猗园景区历史遗存众多，幽谷秀雅。曲溪鹤影景区三面环水，风光宜人。花香仙苑景区视野开阔，俊朗明快。幽篁烟月景区以竹为特色，竹石相映，景色淡雅。作为曾经的文人园林，古猗园的特色有五。一在文人之意境。营造之初，朱三松即以白居易《池上篇》为参照，"有水一池，有竹千竿"，营造"美竹千竿，净绿如拭"的文人理想境界。如今，古时风貌犹能感知，园内遍植各类翠竹，或挺拔端庄，或婀娜多姿，在水天空阔之中，与建筑融为一体。二在山水之构思。逸野堂、小松冈、戏鹅池区域，沿袭早期山水格局，如丹青高手构图，湖荡逶迤，山池影绰，景物森然，层次不尽，将山水画融入造园艺术。自逸野堂东望，碧波涟漪，曲折处幽深绵长，水聚处疏朗开阔，松冈耸峙，山坡上点缀山石，冈外山池影影绰绰，姿态横生，饱含造园者的巧思。三在艺术之展示。古猗园积累了大量的艺术作品，尤其是园中的匾额、楹联、碑刻和砖雕等。数量众多、独具特色，对园内建筑和

古猗园竹景
Bamboo landscape of Guyi Garden

五老峰
Wulaofeng

  五老峰，清乾隆年间（1736—1795）叶锦置，位于逸野堂北侧，由五块形态各异的太湖石和石琴台组成，有龙仙、鹤仙、鹿仙、鸢仙、鹅仙五老操琴吟唱的传说。

Wulaofeng, placed by Ye Jin during the Qianlong reign (1736–1795) of the Qing dynasty, lies north of Yiye Hall. It consists of five distinct Taihu rocks and a stone platform for Chinese zither. According to legend, five immortals — Dragon, Crane, Deer, Kite, and Goose — gathered here to play music and sing, giving rise to the name "Wulao" (Five Elders).

小松冈
Xiaosonggang

小松冈（上海古猗园提供）
Xiaosonggang (Provided by Shanghai Guyi Garden)

　　古典园林中，叠石不可或缺，园中今日仍有小云兜、小松岗及五老峰等。小松冈，始录于沈元禄《古猗园记》，紧靠戏鹅池和竹枝山，为黄石垒建的半岛，三面环水，东临戏鹅池，南接磬折桥，西对鸢飞鱼跃轩，北连白鹤亭。清咸丰三年（1853）小刀会在南翔活动，小松冈是秘密联络点，竹枝山东坡为转运、埋藏枪支弹药的地点。

Rockeries are a key element of classical Chinese gardens, and in Guyi Garden, notable features such as Xiaoyundou, Xiaosonggang, and Wulaofeng remain prominent today. Xiaosonggang, first recorded in Shen Yuanlu's *Record of Guyi Garden*, is located near Xi'e Pond and Zhuzhi Hill. This yellow-stone-built peninsula is surrounded by water on three sides, with Xi'e Pond to the east, Qingzhe Bridge to the south, Yuanfei Yuyue Pavilion to the west, and Baihe Pavilion to the north. During the Taiping Heavenly Kingdom Movement in 1853, Xiaosonggang served as a secret meeting point for the Small Swords Society, while the eastern slope of Zhuzhi Hill was used for transporting and concealing weapons and ammunition.

不系舟，又名"书画舫"，俗称"石舫"，始建于明嘉靖初期，位于戏鹅池北岸，三面临水。舟名出自白居易《适意》："岂无平生志，拘牵不自由。一朝归渭上，泛如不系舟。"建成后明代书法名家、"吴中四才子"之一的祝允明游览猗园，为石舫题"不系舟"额，遂声名远播。"猗园戏鹅""舢舫观鱼"为"南翔八景"中的两景。不系舟历史久远，屡经修缮。今不系舟由楼、廊、亭、阁组合，舟头有锚凳、舱门、漏窗，两舷设上下花窗，后舱建隔门，左右进出，舟尾设梯上下阁楼。舟廊宽敞，凭栏可赏戏鹅池全景。

The Buxizhou (Unmoored Boat), commonly referred to as the Stone Boat, was originally built during the early Jiajing period of the Ming dynasty. Situated on the northern shore of Xi'e Pond, it is surrounded by water on three sides. The name "Buxizhou" was inspired by a poem from the renowned Tang dynasty poet Bai Juyi: "Though dreams I held, they were confined. My steps were bound, no path to find. One day, I returned to the Wei's calm flow. Like an unmoored boat, drifting slow." After its completion, the famous Ming dynasty calligrapher Zhu Yunming, one of the "Four Talents of Suzhou," visited Yi Garden and inscribed the name "Buxizhou" on the boat, giving it widespread fame. The scenes "Watching Geese in Yi Garden" and "Viewing Fish from the Boat" later became two of the "Eight Scenic Views of Nanxiang." With its long history, Buxizhou has undergone numerous restorations. The current structure combines elements of a pavilion, corridor, and loft. At the bow, there are anchor benches, cabin doors, and latticed windows. Both sides feature upper and lower latticed windows, and partition doors lead to the rear cabin with entrances on either side. A staircase at the stern provides access to the loft. The spacious corridors offer visitors panoramic views of Xi'e Pond, allowing them to fully appreciate the tranquil beauty of the surrounding waterscape.

不系舟
Buxizhou

上 不系舟近景
U: Detailed view of Buxizhou (Stone Boat)
下: 不系舟全景
D: Full view of Buxizhou (Stone Boat)

小云兜
Xiaoyundou

　　江南园林以叠石理水为主要景观，古猗园自营造之初，便有多处相关景致。小云兜位于逸野堂南侧，占地约40平方米，高约3米，由明代竹刻家朱三松督造，清代叶锦拓修为兜云洞，因如晨雾中的淡灰云朵而得名。山顶有平台可观景，山内有洞可穿行，后部植松，与之成景。1937年淞沪会战时被日机炸损，1957年、1983年进行了两次修缮。

Rockeries and waterscapes, central elements in Jiangnan gardens, have always played a vital role in defining the scenic beauty of Guyi Garden. Xiaoyundou, located at the south of Yiye Hall, spans about 40 square meters and rises approximately 3 meters high. Originally supervised by the Ming dynasty bamboo carver Zhu Sansong and later expanded by Ye Jin during the Qing dynasty, it was named Douyun Cave for its resemblance to a pale gray cloud floating in the morning mist. The formation includes a viewing platform at the top and a cave that visitors can walk through, with pines planted at the rear to enhance its scenic beauty. Although Xiaoyuntai was damaged in the Japanese bombings during the 1937 Battle of Shanghai, it was restored in 1957 and 1983, preserving its natural beauty for future generations.

洛阳桥
Luoyang Bridge

洛阳桥,建于清乾隆年间(1736—1795),原位于古猗园外东北侧,横跨万四浜,石桥长 6.37 米,宽 1.87 米。该桥因乾隆皇帝下江南而建造,桥板两侧刻有双龙,属御桥,百姓不得通行,遂弃用。1983 年移置龟山西南侧。

Luoyang Bridge, built during the Qianlong reign (1736–1795) of the Qing dynasty, originally spanned Wansi Creek, northeast of Guyi Garden. This stone bridge, measuring 6.37 meters long and 1.87 meters wide, was constructed to commemorate Emperor Qianlong's southern tour. Its deck is carved with double dragon motifs. As an imperial bridge, it was reserved exclusively for the emperor, and commoners were prohibited from crossing it, which eventually led to its disuse. In 1983, the bridge was relocated to the southwest of Guishan District.

翠霭楼
Cui'ai Tower

  翠霭楼,始建于清代,原为旧南市区和顺街火神庙的打唱台,高13米,占地90平方米,全木结构。正面花窗开阔雄伟,楼上中央出一阁,凸面雕饰华丽。二楼藻井精雕双龙戏珠,斗拱精巧如鸟头,谓"百鸟朝凤"。陈从周先生评价其为"典型清代宏伟建筑",1986年移置古猗园翠猗桥东,并建独立庭院,取名"艺苑"。

  Cui'ai Tower, originally built during the Qing dynasty, once served as the opera stage of the Fire God Temple on Heshun Street in the former Nanshi District. Standing 13 meters tall and covering 90 square meters, the tower is constructed entirely of wood. Its front is adorned with grand, latticed windows, while the second floor features a central structure adorned with intricate carvings. The coffered ceiling on the second floor showcases a striking depiction of two dragons playing with a pearl, and the delicate brackets, shaped like bird heads, are said to symbolize "a hundred birds paying homage to the phoenix." Esteemed scholar Chen Congzhou praised it as a "classic grand structure of the Qing dynasty." In 1986, the tower was relocated to the east of Cuiyi Bridge in Guyi Garden, where a dedicated courtyard, now known as Yi Garden, was built around it to preserve its cultural significance.

万安塔
Wan'an Pagoda

　　万安塔，原位于南翔镇元代万安寺内，为五层楼阁式实心石塔，通高约3.5米，底面为八角形，基座为须弥座，底层每面雕有壸门，二至五层每面刻有佛龛与佛像。各层均有腰檐，塔顶由八角形塔盖、石鼓、仰覆莲座、塔刹等组成。塔檐部刻有信徒姓氏、生辰及捐款数额等。1937年淞沪会战中寺毁，万安塔受损严重，埋于土中，1988年移置青清园门庭西侧。

One such structure is the Wan'an Pagoda, originally located at Wan'an Temple in Nanxiang during the Yuan dynasty (1271–1368). This five-story solid stone pagoda stands about 3.5 meters tall, with an octagonal base resting on a Mount Sumeru platform. The first tier features a cusped gable doorway on each side, while the second to fifth tiers are embellished with Buddha niches and seated Buddha figures. Each tier is capped with waist eaves, and the top of the pagoda includes an octagonal roof, a stone drum, a double lotus throne, and a finial. Inscriptions on the eaves record the names, birthdates, and donation amounts of the pagoda's patrons. The temple was destroyed during the Battle of Shanghai in 1937, and the Wan'an Pagoda was heavily damaged and buried. In 1988, the pagoda was relocated to the west of Qingqing Garden's gate.

微音阁
Weiyin Belvedere

  微音阁，建于 1947 年，坐北朝南，砖木结构，楼阁式两层建筑，错落典雅。该建筑由南翔进步青年组织微音社募捐建成，是中共党组织在南翔的活动场所之一。微音，意指在黑暗社会发出正义呐喊，点燃民众向往和平光明的星星之火。微音虽小，但精神长存。

  Weiyin Belvedere, built in 1947 during the Republic of China period, is a two-story brick-and-wood structure facing south, with an elegant, staggered design. It was funded through donations raised by the "Weiyin Society," a progressive youth group in Nanxiang, and served as one of the local activity centers for the Communist Party. The name "Weiyin" symbolizes a call for justice in a dark society, inspiring hope for peace and light among the people. Though its voice was faint, its spirit endured.

补阙亭
Buque Pavilion

左:补阙亭
L: Buque Pavilion
右:补阙亭内景
R: Interior of Buque Pavilion

补阙亭,又名"缺角亭",位于古猗园竹枝山,1931年"九一八"事变后东北三省被日军侵占,为铭记国耻,南翔人民特建此亭。补阙亭为砖木结构,方形四角攒尖顶,以四柱支承。东北角残缺,表示东北沦陷,另三角作握拳状,满腔的义愤通过三只高高举起的拳头展现,表达了人民收复失地、昭雪耻辱的决心,为园林留下一段抗战的不朽史诗。

The Buque Pavilion, also known as the "Pavilion of the Missing Corner," stands on Zhuzhi Hill in Guyi Garden. After the Japanese occupation of China's northeastern provinces following the September 18th Incident (1931), the people of Nanxiang constructed this pavilion to commemorate the national disgrace. The pavilion is a brick-and-wood structure with a square base, a four-sided pyramidal roof, and four supporting pillars. Its northeastern corner is deliberately left incomplete, symbolizing the loss of the northeastern provinces. The other three corners are shaped like raised fists, representing the people's determination to reclaim the lost land and restore national honor. The pavilion stands as a powerful symbol of China's resistance to Japanese aggression and an enduring testament to the garden's wartime history.

普同塔
Putong Pagoda

普同塔远景
Distant view of Putong Pagoda

普同塔，建于南宋嘉定十五年（1222），同为南翔寺遗物，塔身青石质地，六面三级，由基座、塔身、宝顶构成。第一级塔身一面开龛，内雕佛像，旁有铭文，记述塔名、建造及重修年代。第二级塔身每面造像，有立姿与坐姿菩萨及行脚比丘像。第三级塔身浮雕云纹。每层塔身上置仰覆莲腰檐。上为八角形攒尖顶，雕出瓦陇，再上为仰莲与宝瓶构成的塔刹。1959年移置古猗园，1987年迁入园中荷花池内。

The Putong Pagoda, built in 1222 during the Song dynasty's Jiading reign, is another historical artifact from Nanxiang Temple. This six-sided, three-tiered pagoda is made of greystone and consists of a base, body, and finial. The first tier contains a niche with Buddhas, accompanied by inscriptions detailing the pagoda's name, construction, and restoration dates. The second tier features carvings of standing and seated Bodhisattvas and traveling monks, while the third tier is adorned with cloud motifs. Each tier is capped with double lotus waist eaves. Above the eaves, an octagonal pyramidal roof, carved to resemble tile ridges, is topped with a finial composed of an upturned lotus and a sacred vase. The pagoda was relocated to Guyi Garden in 1959 and moved to its current location in the lotus pond in 1987.

南厅
South Hall

经幢旁的南厅，始建于明代。明崇祯年间（1628—1644），在此发生了奴仆向李氏索还身契的血案，李氏一家数十人遇难，故又称"难厅"，原建筑于1932年"一·二八"事变中被毁，20世纪40年代重建。今南厅坐西北朝东南，砖木结构，面阔三间，进深二间，五架梁，歇山顶，檐下施斗拱，前部有轩廊。

Next to the Dhāraṇī pillar stands the South Hall, originally built during the Ming dynasty. In the Chongzhen period (1628–1644), a tragic event occurred here when servants, demanding the return of their contracts from the Li family, led to the deaths of most family members. This gave the hall its somber nickname, Nan Hall (Hall of Tragedy). The original structure was destroyed during the January 28 Incident (1932) but was rebuilt in the 1940s. The current South Hall, facing southeast, is a brick-and-wood structure with three bays wide and two bays deep. It features a five-beam frame with a double-eaved hip-gable roof, supported by *dougong* brackets under the eaves, and a veranda extending from the front.

古猗园最古之物，为营造于唐代的尊胜陀罗尼经幢。经幢共两座，为南翔寺旧物，曾对称立于大雄宝殿之前，由里人莫少卿捐建，均为青石质地、七级八面，造型壮丽挺秀，镌刻精美。幢身镌有尊胜陀罗尼经文及卷云、莲瓣等纹饰，幢顶镌有狮首和四天王像。北宋太平兴国五年（980）、元元统二年（1334）、清嘉庆中叶先后重修。清末寺废，1949年解放时已倒塌。1959年移置古猗园。一座在古猗园内微音阁南侧，为唐乾符二年（875）建，高10余米。一座在古猗园内南厅北侧，为唐咸通八年（867）建，幢身剥蚀严重，顶上宝刹于1969年遭雷击毁。

The oldest artifacts in Guyi Garden are two Dhāraṇī pillar, also called sutra pillars, from the Tang dynasty, originally part of Nanxiang Temple. Donated by local philanthropist Mo Shaoqing, these greystone pillars, each seven stories tall with eight sides, once stood symmetrically in front of the Mahavira Hall. Adorned with intricate carvings, they feature Uṣṇīṣa Vijaya Dhāraṇī Sūtra inscriptions, cloud scrolls, and lotus petals. The tops are decorated with lion heads and images of the Four Heavenly Kings. The pillars were restored several times — in 980 (Northern Song dynasty), 1334 (Yuan dynasty), and during the mid-Jiaqing period of the Qing dynasty. By the late Qing dynasty, the temple had been abandoned, and by 1949, the pillars had collapsed. They were relocated to Guyi Garden in 1959. One pillar, situated south of Weiyin Belvedere, was built in 875 and stands over 10 meters tall. The other pillar, now located north of the South Hall, was built in 867 during the Tang dynasty but has suffered significant erosion, with its spire destroyed by lightning in 1969.

唐经幢（南厅北侧）
Dhāraṇī pillar of the Tang dynasty (North of the South Hall)

唐经幢
Dhāraṇī Pillar of the Tang Dynasty

唐经幢(微音阁南侧)
Dhāraṇī pillar of the Tang dynasty
(South of Weiyin Belvedere)

园景撷趣
Garden Scenery

古猗园南厅
South Hall, Guyi Garden

　　500年间沧桑变革,凝固成古猗园深厚的历史,除了著录于史志与诗词之中的人物与史迹,还有众多的文物古迹保留在园内。

　Guyi Garden, with its 500-year history, has undergone significant transformations, leaving behind a rich legacy of cultural artifacts and heritage sites. These, along with the historical figures and events commemorated in texts and poetry, are still preserved within the garden today.

such rebellion took place at Yi Garden, where Li Yizhi's three sons, along with Li Liufang's son Li Hangzhi, were tragically killed. Li Yizhi, who was in Nanjing at the time, narrowly escaped. Upon returning to his hometown, he sought refuge at the Hou family's villa in Jiading County, where he lived in seclusion until his death.

As the family's fortunes declined, Yi Garden fell into decline. It passed through the Lu and Li families before being acquired by Ye Jin, a member of the influential Dongting Merchants, in 1746 during the Qianlong reign (1736–1795) of the Qing dynasty. Ye undertook extensive renovations, expanding the garden and redesigning its landscape. He enlisted bamboo carver and artist Zhou Hao to assist with the redesign. These efforts established the layout that defined Guyi Garden until the Republic of China period. Upon completion, the garden was renamed "Guyi Garden" (Ancient Yi Garden) to honor its history spanning two dynasties.

In 1775, as Ye Jin's family fortunes declined, Guyi Garden fell into neglect and became overgrown. By 1788, the local gentry raised funds to purchase much of the garden, donating it to the City God Temple of Taicang as the Sacred Garden. During this time, the Flower God Temple was built, along with statues of the Flower God. In 1806, the gentry again pooled resources to restore the garden.

However, in 1860, during the Taiping Heavenly Kingdom Movement, clashes between Taiping forces, Qing troops, and the Shanghai Volunteer Corps — a small imperial force trained in European tactics — resulted in the destruction of several structures within the garden.

In 1868, Guyi Garden was renovated and opened as a public park. During the Guangxu period (1875–1908), local guilds in Nanxiang added new halls and structures within the garden, using them for meetings and social gatherings. Restaurants, teahouses, and snack shops were also established, allowing visitors to enjoy tea and steamed dumplings in scenic spots such as Yuanyang Hall and Buxizhou (Stone Boat).

In 1901, the "Nanxiang Guyi Garden Liquor Guild" was established in the garden's northwestern corner. By 1908, with the opening of Nanxiang Railway Station along the Shanghai-Nanjing Railway, Guyi Garden became a popular destination for suburban outings, attracting large numbers of visitors. As urbanization progressed, classical gardens like Guyi Garden gradually evolved into public parks and popular tourist attractions.

In 1932, during the January 28th Incident, Guyi Garden was occupied by Japanese forces for over two months, causing significant damage. After their retreat, local patriots raised funds to restore the garden. However, in 1937, during the Battle of Shanghai, Guyi Garden was bombed by Japanese aircraft, and most of its structures were destroyed. Beginning in 1946, parts of the garden were rebuilt, including the construction of Weiyin Belvedere.

After 1949, Guyi Garden was restored and reopened to the public on October 1, 1959. Since then, the garden has undergone several renovations and expansions, now covering approximately 98 200 square meters. It is divided into four scenic areas: Yi Garden (the original garden), Quxi Heying, Huaxiang Xianyuan, and Youhuang Yanyue. The garden's intricate design, distinctive landscapes, and ancient cultural artifacts beautifully revive the historical charm and elegance of this treasured site.

园内翠竹
Emerald bamboo in the Garden

古樹閑逸野堂
Tiye Hall, Guyi Garden

Guyi Garden was originally constructed approximately between 1522 and 1526 by the Min family from Huizhou and was first named "Jie Garden." The garden's owners were Min Shanglian, and his son, Min Shiji (style name Mingqing), who served as Director of the Court of Imperial Entertainments and Assistant Prefectural Magistrate of Song County in Henan. During the Wanli period (1573–1620) of the Ming dynasty, Min Shiji invited Zhu Zhizheng, a renowned bamboo carver, calligrapher, and painter from Jiading, to design the garden. Zhu transformed it into "a residence of 10 *mu* and a garden of 5 *mu*," featuring "a pond and a thousand bamboo stalks." The garden was renamed "Yi Garden," with "Yi" meaning lush bamboo, inspired by a verse from *The Book of Songs*, "Look at those recesses in the banks of the Qi River, with their green bamboo, so fresh and luxuriant."

In the early Chongzhen period (1628–1644) of the Ming dynasty, Yi Garden was acquired by Li Yizhi, a member from a family with a rich scholarly tradition and a profound appreciation for gardens. Adopting the name "Yuyuan Jushi" (Layman of the Garden Residence), Li redesigned Yi Garden, imbuing it with the refined tastes of the literati. He recorded his reflections in *Record of Life in the Garden*. Yi Garden became one of the "Three Gardens of the Li Family," serving as a cultural hub for gatherings, literary exchanges, and strolls, leaving a lasting legacy of poetry and literary works.

In 1644, during the chaotic final days of the Ming and Qing dynasty alternation, bondservant uprisings swept across the Jiangnan region. Impoverished farmers banded together to attack the estates of their masters, freeing servants and destroying tenant contracts. One

"白鹤南翔"传说来历诗文碑
Stele of the legend and poems of the *White Crane Flying South*

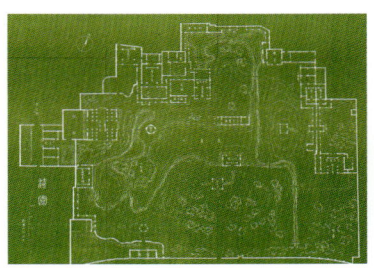

《江南园林志》古猗园平面图(20世纪30年代童寯先生绘)
Plan of Guyi Garden, Gardens of Jiangnan
(Painted by Tong Jun in the 1930s)

1959年重建后的不系舟(上海古猗园提供)
Buxizhou after Reconstruction in 1959
(Provided by Shanghai Guyi Garden)

清咸丰十年（1860）太平天国运动中，太平军在南翔与清军及"洋枪队"展开激战，园内部分建筑被毁。

清同治七年（1868）重修园林，改为公园。至光绪年间（1875—1908），南翔各同业公所在园内修厅堂、增建筑，作为集议场所，开设酒楼、茶肆、点心店，游人可在鸳鸯厅、不系舟叙茶、品小笼。

清光绪二十七年（1901），园西北建"南翔古猗园酒业公所"。光绪三十四年（1908），沪宁铁路建南翔火车站，古猗园成为上海郊游之地，大量游客前来游览。古典园林在近代城市化浪潮中，逐渐转换为近代公园、郊游之地。

1932年"一·二八"事变期间，古猗园被日军侵占并盘踞达两月有余，园景多遭破坏。日军撤退后，当地爱国人士筹资对其进行了修复。1937年淞沪会战，古猗园又遭日机轰炸，大部分建筑毁于战火。1946年后，古猗园部分重建，并新建微音阁。

1949年后，古猗园经过修缮于1959年10月1日正式对公众开放，之后又历经多次改扩建，现古猗园总面积约98 200平方米，按不同景观划分为猗园（老园）、曲溪鹤影、花香仙苑、幽篁烟月四个景区，精巧的艺术构思、独特的园林景观与多处古朴的文物古迹，重现了古园的历史风貌和秀丽神韵。

20 世纪初的不系舟（上海古猗园提供）
Buxizhou in the Early 20th Century (Provided by Shanghai Guyi Garden)

1911 年的不系舟（上海古猗园提供）
Buxizhou of 1911 (Provided by Shanghai Guyi Garden)

20 世纪 30 年代的不系舟（上海古猗园提供）
Buxizhou in the 1930s (Provided by Shanghai Guyi Garden)

20 世纪 30 年代小松冈及周边景象（上海古猗园提供）
Xiaosonggang and Surrounding Scenery in the 1930s
(Provided by Shanghai Guyi Garden)

随着家族的衰落，园林也未曾幸免。猗园之后转让给陆姓、李姓人家。清乾隆十一年（1746），洞庭山人叶锦购得猗园，大兴土木、修葺装点，拓充园基，改筑园景，并延请竹刻家、书画家周颢参与设计，从而奠定了古猗园直至民国时期的基本格局。竣工后，因隔了一个朝代，更名为"古猗园"。

乾隆四十年（1775），叶锦家道中落，园林失修，逐渐荒芜。乾隆五十三年（1788），地方士绅募购古猗园大部分区域，捐给太仓州城隍庙作为灵苑。其间，园内募建花神庙、塑花神像等。清嘉庆十一年（1806），地方乡绅又集资进行修缮。

历史沿革

Garden History

古猗园始建于明嘉靖初年(具体建造时间大致为 1522—1526 年),由徽籍闵氏建造,初名"猗园"。园主闵尚廉,其子闵士籍,字明卿,官至光禄寺署正、河南嵩县通判。明万历年间(1573—1620),闵士籍邀嘉定著名竹刻家、书画家朱稚征(号三松)擘画设计园林。朱稚征为"嘉定三朱"之一,工诗善画,尤擅竹刻,猗园经其构思,以"十亩之宅,五亩之园"营构,园中"有水一池,有竹千竿",取《诗经·卫风·淇奥》中"瞻彼淇奥,绿竹猗猗"之意,更名为"猗园"。

明崇祯(1628—1644)初年,猗园转让给"嘉定四先生"之一、书画家李流芳之侄、贡生李宜之。李宜之为翰林李名芳之子,酷爱园林,号"寓园居士",其以书画家的布局、文人特有的趣味,修造猗园,并撰《园居》记之。猗园也成为"李氏三园"之一[2],承载了游宴雅集等社交活动,留下大量诗词歌赋等文学作品[3]。

崇祯十七年(1644),明清鼎革之际,江南奴变频发,猗园发生家奴索契事件,李宜之三子和李流芳之子李杭之等遇难。宜之身在南京而躲过一劫,回乡后寄居在嘉定县城侯氏别业中,直到去世[4]。

《嘉定县志》古猗园图,清嘉庆十六年(1811)
Guyi Garden Map, Gazetteer of Jiading County, 1811

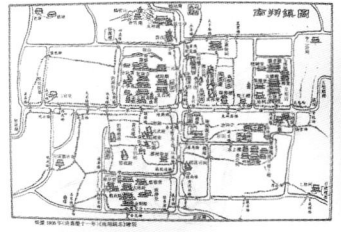

《南翔镇志》南翔镇图,清嘉庆十一年(1806)
Map of Nanxiang Town, Gazetteer of Nanxiang Town, 1806

曲香廊
Quxiang Corridor

古猗园坐落于中国历史文化名镇南翔。南梁天监年间（502—519）此地创建南翔寺，后因寺成镇，以寺得名。至宋元时期，已颇为繁华。明代以来，商贾辐辏、百货俱集，人文荟萃。经济繁荣促进了文化的发展，南翔镇人才辈出，涌现出一批知名学者、文化名流，《南翔镇志》载"明代全镇有进士十人、举人十六人，贡生十四人"[1]。在经济与文化的丰厚积淀下，造园风气席卷南翔，聚居此地的官宦世家、富商大贾、文人雅士竞相兴建宅第园林，因此南翔也是园林聚集之地。"小小南翔赛苏城"，史志记载南翔曾建园十余座，营建之势，蔚为壮观。古猗园即诞生于其间，是南翔镇遗留至今的园林明珠。

Guyi Garden is located in the Famous Historical and Cultural Town of Nanxiang. The town traces its origins to the Tianjian period of the Liang dynasty (502–519), when Nanxiang Temple was established, and it later took its name from the temple. By the Song and Yuan dynasties, it had already become quite prosperous. From the Ming dynasty onward, Nanxiang became a thriving center of commerce and culture, attracting merchants and scholars alike. According to the town gazetteer, "In the Ming dynasty, ten *jinshi* (imperial palace examination graduates), sixteen *juren* (provincial imperial examination graduates), and fourteen *gongsheng* (tribute students selected from local academies) hailed from Nanxiang." Under the town's rich economic and cultural foundation, a wave of garden construction spread across Nanxiang. Prominent families, wealthy merchants, and scholars in Nanxiang eagerly built grand estates and gardens, making the town a hub of remarkable garden design. Historical records describe Nanxiang as "a small town rivaling Suzhou," with more than 10 gardens at its height, creating a spectacular landscape. Among these, Guyi Garden, a gem from this flourishing era, remains one of the most cherished examples of Nanxiang's historic gardens.

南苑夏景一隅
A corner of the summer scenery at South Garden

猗猗亭
Yiyi Pavilion

古猗园
Guyi Garden

曲水潆洄，海上秀园——古猗园
Flowing Waters and Timeless Charm: Guyi Garden

年代：明代
地址：上海市嘉定区南翔镇沪宜公路 218 号
占地面积：约 98 200 平方米
保护级别：上海市文物保护单位

Era: Ming dynasty
Address: No.218, Huyi Highway, Nanxiang Town, Jiading District, Shanghai
Area: Approximately 98 200 square meters
Protection level: Shanghai city-level protected site

古猗园鸟瞰图（上海古猗园提供）
Aerial view of Guyi Garden (Provided by Shanghai Guyi Garden)

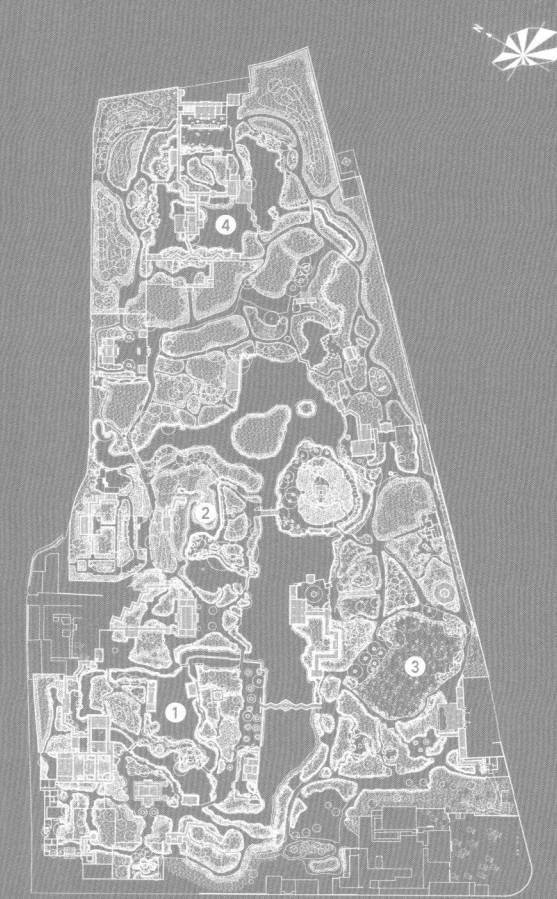

图例
Legends

1 倚园景区
  Yi Garden Scenic Area
2 曲溪鹤影景区
  Quxi Heying Scenic Area
3 花香仙苑景区
  Huaxiang Xianyuan Scenic Area
4 幽篁烟月景区
  Youhuang Yanyue Scenic Area

古猗园

Guyi Garden

观水亭
Guanshui Pavilion

闲研斋与致雨斋
Xianyan Studio and Shuyu Studio

艺文丰韵
Cultural and Artistic Legacy

秋霞圃历经520多年的岁月沧桑，园内文人雅士留下的雪泥鸿爪多不胜数，它们既是秋霞圃的人文瑰宝，也是研究秋霞圃与嘉定历史的珍贵资料。据《秋霞圃志》所载，秋霞圃中有匾额46块、楹联91副、碑刻38件、墓志4篇；与之相关的题词23首、书画118件；另有与秋霞圃相关的诗词文章、影视作品不胜枚举。这些艺文瑰宝共同丰富着秋霞圃的内涵与底蕴。

With a history spanning over 520 years, Qiuxia Garden has long been a gathering place for literati, leaving behind a rich cultural heritage. These cultural treasures not only form a vital part of the garden's legacy but also offer valuable insights into the history of both Qiuxia Garden and Jiading District. According to *Records of Qiuxia Garden*, the garden houses 46 plaques, 91 couplets, 38 steles, and four epitaphs. Additionally, there are 23 inscriptions, 118 pieces of calligraphy and paintings, as well as numerous poems, articles, and films associated with the garden. Collectively, these cultural treasures enrich the legacy and richness of Qiuxia Garden, making it a trove of historical and artistic significance.

1621年，明代书法家娄坚题"涉趣桥"（1838年复刻，上海市嘉定博物馆提供）
注：桥位于桃花潭南岸，明末属沈氏园
Inscription of "Shequ Qiao" (Scenic Bridge, Lou Jian, Ming dynasty, carved in 1621 and re-carved in 1838, provided by Shanghai Jiading Museum)
Note: the bridge, located on the southern bank of Taohua Pool, belonged to the Shen Family Garden in the late Ming dynasty

> **参观指南 Visiting Guide**
>
> 开放时间：08:00—16:30。
> 交通信息：地铁11号线至嘉定北站，换乘嘉定12路至博乐广场站。
> Opening hours: 08:00-16:30.
> Traffic information: Metro Line 11 to North Jiading Station, transfer to Jiading Bus 12 to Bole Square Station.

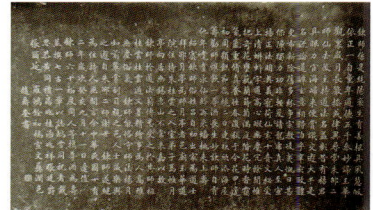

《练师吕绍宾六十寿庆记》碑文，1923（上海市嘉定博物馆提供）
Inscription of "Celebrating Taoist Master Lü Shaobin's 60th Anniversary," 1923 (Provided by Shanghai Jiading Museum)

之比为1∶1.77，常绿树与落叶树数量之比为1∶0.6。园内现有银杏、枸骨、榉树等古树名木12株，后续资源21株。这些古树名木或树高冠大、或树形古拙，皆绿荫婆娑、灵气十足，成为秋霞圃园林组景的主要景观。

值得一提的是，在秋霞圃，除了秋赏红枫，春天亦是百花争艳，尤以牡丹国色艳压群芳。近年来，园方连续举办数届牡丹展，游人如织，声名远播。以2024年为例，在花展上，人们不仅可于清境塘西侧牡丹园内领略百年牡丹的风采，还可欣赏岛锦、彤云等28种、300余株中国传统精品牡丹，娇艳繁花与古典园林相得益彰。

Qiuxia Garden is renowned for its autumn beauty. Each year, in late autumn, the garden transforms into a vibrant tapestry of colors, adding warmth to the cool days. Fiery red maple leaves, golden ginkgo trees, lush green bamboo, and dark green pines blend into a dazzling display, reminiscent of brilliant clouds. As Tang poet Liu Yuxi wrote, "Autumn was once lamented for its desolation, but I believe autumn surpasses the splendor of spring." A stroll through Qiuxia Garden in this season reveals a landscape where even the falling leaves contribute to a masterpiece of color. Wandering through the garden's autumn scenery, visitors become part of the landscape, immersed in its beauty.

Qiuxia Garden is home to an abundance of ancient trees and renowned plants, with lush greenery and flowers blooming throughout the year. These plants, in harmony with the garden's architecture, landscapes, and aesthetic design, create an artistic realm that is both poetic and full of life. This balance has earned Qiuxia Garden the reputation of being "man-made, yet as natural as heaven's creation." As of 2000, the garden is home to 3964 trees, 209 herbaceous plants, 49 vines, and 2405 bamboo plants, including over 300 sacred bamboos, a signature feature of the Jiangnan gardens. The ratio of trees to shrubs is 1∶1.77, and the ratio of evergreen to deciduous trees is 1∶0.6. Among these, 12 ancient trees, such as ginkgo, holly, and zelkova, stand as proud symbols of the garden's heritage, with another 21 notable tree species. These ancient trees, some towering and majestic, others gnarled and rustic, create a rich, shaded environment that plays a vital role in shaping Qiuxia Garden's scenic beauty.

Qiuxia Garden is not only a spectacle of red maples in autumn but also a vibrant showcase of blooming flowers in spring, with peonies reigning supreme. In recent years, the garden has hosted popular peony exhibitions, drawing large crowds and widespread attention. In 2024, for example, visitors admired century-old peonies in the Peony Garden on the west side of Qingjing Pond, alongside 28 varieties of traditional Chinese peonies, with over 300 plants in bloom. The vibrant flowers perfectly complement the classical garden setting.

秋霞"枫"景

Autumn Scenery in Qiuxia Garden

秋霞圃以秋景闻名,"秋霞"两字道出此园的最美时节。每年深秋,五彩缤纷的植被交相辉映,给清冷的秋日增添了几抹浓烈的暖色。枫叶红、银杏黄、修竹翠、松柏青……灿若云霞、风情万种。"自古逢秋悲寂寥,我言秋日胜春朝",在秋霞圃中漫步,就连落叶也描绘着层林尽染的意境。深秋的秋霞圃"枫"景如画,人们徜徉画中,便也成了风景,一切都是那么美好。

秋霞圃古树名木资源丰富,繁茂葱茏间亦有四时花木争奇斗艳。这些美丽的植物与园内建筑、景致、人文、美学等意趣和谐统一,营造出一种诗情画意、灵动脱俗的艺术境界,使秋霞圃赢得了"虽由人作,宛若天开"的美誉。据 2000 年的统计数据,全园共有树木 3964 棵,草本花卉 209 株,藤本 49 株,竹 2405 株,其中富有江南园林特色的南天竹 300 余株。乔木与灌木数量

秋霞圃的红枫
Red maples in Qiuxia Garden

井亭
Well Pavilions

寝宫为南向,三楹七架,卷棚顶,上覆小青瓦屋面,以50厘米见方的方砖铺地,这种规格的地砖与嘉定孔庙明伦堂(重建于1876年)为同一尺寸,明伦堂重建年代与城隍庙仅相隔6年,故该方砖很有可能是清代所制。

井亭位于秋霞圃东大门外,两亭分立大门两侧,为原邑庙遗物,1986年修葺。两亭皆为正方形,四角立四根方石柱,四面以石坐凳围之,两亭相向留一个进出口。亭内置六角形古井,地面以城砖侧铺成棋盘图案,较为罕见。

The residential palace faces south, with three bays and seven purlin beams, featuring a rounded main ridge and a roof covered in small gray tiles. The floor is paved with 50-centimeter square bricks, the same size as those used in the Minglun Hall of the Temple of Confucius in Jiading, which was rebuilt in 1876. Since the rebuilding of both Minglun Hall and the City God Temple occurred only six years apart, it is likely that these bricks date from the Qing dynasty.

The Well Pavilions are located outside the East Gate of Qiuxia Garden, standing on either side of the entrance. These structures are remnants of the original City God Temple and were restored in 1986. Each square pavilion is supported by four stone pillars at the corners, with stone benches lining all four sides. The pavilions face each other, with an entrance on each side. Inside, a hexagonal ancient well is set, and the floor is paved in a unique checkerboard pattern using bricks.

城隍庙景区以大殿为中心，西侧有月门连通桃花潭景区的聊淹堂，东侧有石板路连接凝霞阁景区和清镜塘景区。

城隍庙大殿于明洪武三年（1370）自南城富安坊移建至今址，明、清时期因火灾和兵祸屡毁屡建,清光绪八年(1882)重建,今日大殿及14米工字廊、寝宫均为当时遗存。1983年，大殿按原样大修，现为重檐歇山顶，双顶连体。前殿三楹十架，后殿五楹十一架。上覆筒瓦，暗亮螭吻脊，脊中塑盘龙戏珠。垂脊端部塑"八仙"坐像，钉帽檐口，龙寿瓦当，歇山墙置如意纹饰裹衣板。大门设直棂格落地长窗20扇，后殿左右两侧各置花格半窗6扇。殿前接双脊檩翻轩明廊，殿北接工字廊与寝宫。

At the heart of the City God Temple scenic area stands the main hall, with a moon gate on the west connecting to Liaoyan Hall in the Taohua Pool scenic area and a stone path on the east linking to the Ningxia Belvedere and Qingjing Pond scenic areas.

The main hall was originally relocated to its current site from Fu'an Ward in the southern city in 1370, during the Ming dynasty's Hongwu reign. Over the centuries, the temple was repeatedly destroyed and rebuilt due to fires and wars, with the last major reconstruction taking place in 1882 during the Qing dynasty. The current main hall, along with the 14-meter-long I-shaped corridor and residential palace, dates back to that time. In 1983, the hall was restored to its original form. It features a double-eave hip-gable roof. The front hall features three bays and 10 purlin beams, while the rear hall features five bays and 11 purlin beams. The roof is adorned with semicylindrical tiles and ridge ornaments, with dragons playing with a pearl at the center of the main ridge. Statues of the Eight Immortals decorate the ends of the ridges framing the gable, while cloud-shaped ornaments line the walls. The main entrance features 20 long vertical windows with intricate lattice patterns, and the rear hall has six latticed half-windows on either side. The hall is connected to a double-ridged corridor at the front and to the I-shaped corridor leading to the residential palace at the back.

城隍庙景区
City God Temple Scenic Area

三隐堂
Sanyin Hall

  每年 12 月中旬,清镜塘景区内的百棵枫树迎来最佳的赏红时节,满园的红枫如红霞栖落,恰似杜牧《山行》诗中那般浓郁绚烂,"霜叶红于二月花"的美景着实令人震撼。红枫的周边还种植了金黄色的银杏、红色的榉树、橙色的乌桕,届时它们也会呈现出油画般饱满的色度,与红枫高低错落、远近相宜,在阳光的照射下,五彩斑斓、熠熠生辉。

  In mid-December, the hundred maple trees in the Qingjing Pond scenic area reach their peak, draping the garden in a stunning display of red leaves, reminiscent of rosy clouds. The sight evokes the beauty described in *A Mountain Walk*, a poem by Tang dynasty poet Du Mu, where "frost-bitten leaves look redder than early spring flowers." The red maples are complemented by golden ginkgoes, red zelkovas, and orange Chinese tallow trees, creating a vivid, painterly landscape. Under the sunlight, the colors — red, gold, and orange — come alive, glowing with radiant brilliance.

三隐堂位于清镜塘东北岸，建于1985年，部分建造材料取自嘉定南门涛阁，并移用了龚弘旧居三隐堂之名，由陈从周题额。

清镜塘的西部为枫树岭，原名青松岭，为1985年挖塘堆土而成，遍植枫树、青松等树木。西南坡有黄石堆叠的假山一座，雄奇陡峻，原有瀑布，现瀑布已废弃。假山下有一湾小溪，溪中设有石矶连通两岸。枫树岭山顶有一座长方形的亭子，名为"岁寒亭"，亭内有一个方形石台和四个石凳，购于城内花园弄的一处民居。亭的四周栽有松、竹、梅，取"岁寒三友"之意。

Sanyin Hall, located on the northeastern shore of Qingjing Pond, was constructed in 1985 using materials from Tao Belvedere at Jiading's South Gate. The hall's name was borrowed from Gong Hong's original residence, and its plaque was inscribed by Chen Congzhou.

To the west of Qingjing Pond lies Maple Ridge, formerly known as Pine Ridge. Formed in 1985 from excavated pond soil, the ridge is covered with maples and pines. On the southwestern slope, a steep and majestic yellow-stone rockery once featured a waterfall, which is now no longer in use. A small creek flows beneath the rockery, with stone jetties connecting both banks. At the summit of Maple Ridge stands Suihan Pavilion, a rectangular structure housing a square stone table and four stone stools, originally sourced from a private residence on Huayuan Lane. The pavilion is surrounded by pine, bamboo, and plum trees, symbolizing the "Three Friends of Winter."

枫树岭
Maple Ridge

岁寒亭
Suihan Pavilion

清镜塘景区
Qingjing Pond Scenic Area

　　清镜塘景区为金氏园遗址，有三隐堂、秋水轩、枫树岭等景点，景区内植物繁盛、疏朗有致，颇具山野之趣。

　　清镜塘原是嘉定城内练祁河的支流，1974年被填平作为启良学校操场的一部分，1985年重疏成塘。塘的东西北三端为水面开阔的池塘，其余水面狭窄如溪，岸线曲折多变，并有河道与桃花潭相通。清镜塘东面宽阔处有一绿岛，从东至西有观荷、绿荫、听松三座石板平桥，北门有清镜桥。

The Qingjing Pond scenic area, once part of the Jin Family Garden, includes notable features such as Sanyin Hall, Qiushui Gallery, and Maple Ridge. This area is lush with greenery, creating a natural atmosphere.

Originally a branch of Jiading's Lianqi River, Qingjing Pond was filled in 1974 to become part of Qiliang School's playground. In 1985, it was dredged and restored to its original form. The eastern, western, and northern sections of the pond are wide and open, while the remaining water narrows into stream-like channels with winding, irregular shorelines. The pond also connects to Taohua Pool via a waterway. A green island sits on the wide eastern section, and from east to west, three stone slab bridges — Guanhe Bridge, Lüyin Bridge, and Tingsong Bridge — cross the water. Visitors will also find Qingjing Bridge at the northern gate.

凝霞阁的东南方为环翠轩，1920年重建，1985年翻建。环翠轩为三楹七架，东侧回廊北连洗句亭，南接扶疏堂。轩内葱翠茂盛，前院有一口古井，井口石栏圈呈六角形，镌有正楷阳文"义井"。

Southeast of Ningxia Belvedere is Huancui Gallery, rebuilt in 1920 and renovated in 1985. The gallery features three bays and seven purlin beams, with an eastern corridor connecting it to Xiju Pavilion in the north and Fushu Hall in the south. Surrounded by lush greenery, the gallery's front courtyard contains an ancient well, featuring a hexagonal stone wellhead inscribed with the words "Yi Jing" (Righteous Well) in raised regular script.

环翠轩
Huancui Gallery

屏山堂与宾藻风香室
Pingshan Hall and Binzao Fengxiang Room

20 世纪 50 年代的屏山堂
（上海翥云艺术博物馆藏，上海市嘉定博物馆提供）
Pingshan Hall, 1950s (Shanghai Zhuyun Art Museum Collection, provided by Shanghai Jiading Museum)

20 世纪 70 年代的屏山堂
（上海翥云艺术博物馆藏，上海市嘉定博物馆提供）
Pingshan Hall, 1970s (Shanghai Zhuyun Art Museum Collection, provided by Shanghai Jiading Museum)

凝霞阁，又名"迎霞阁"，1921年重建时为五楹两层阁，1985年改建为三楹六架，西楹上有阁，阁额由宋日昌题写。阁的四周有回廊，阁前庭院宽敞，有湖石假山且绿树成荫，是登阁俯瞰桃花潭和清镜塘的最佳处。

凝霞阁西南侧是屏山堂与宾藻风香室，此处堂室连为一体，呈"凸"字形，建于1921年。朝东的三楹为屏山堂，朝西的一楹为宾藻风香室，中间有砖墙相隔。屏山堂因堂前湖石假山得名，假山高约3米，石质坚实润泽，形似屏风，雅趣天成。堂前地势低洼，采用"旱园水做"的假象法，三面以湖石堆砌，给人们以"池"的感觉。堂前湖石西向的半亭即为宾藻风香室，室内置屏门，两侧磨砖坐槛，南向轩廊接闲研斋。堂的南山墙处种有一株130余年的山茶树。

Ningxia Belvedere, also known as Yingxia Belvedere, was originally a five-bay, two-story structure rebuilt in 1921. In 1985, it was renovated into a three-bay, six-purlin-beam structure, with an inscription on the western bay by Song Richang, the former Deputy Mayor of Shanghai. Encircled by corridors, the belvedere opens to a spacious front courtyard featuring a rockery made of lake rocks and shaded by lush trees. This scenic spot offers sweeping views of both Taohua Pool and Qingjing Pond, making it an ideal location to appreciate the garden's natural beauty.

To the southwest of Ningxia Belvedere stands an inverted T-shaped structure built in 1921, comprising Pingshan Hall and Binzao Fengxiang Room. Pingshan Hall occupies the three eastern bays, while Binzao Fengxiang Room takes up the single western bay, separated by a brick wall. Pingshan Hall, meaning "Hall of the Rockery Screen," is named after the Dapingshan Rockery in front, which stands about 3 meters tall and is made of solid, smooth lake rocks, resembling a natural screen. The slightly sunken ground in front creates the illusion of a water feature, with lake rocks piled on three sides to mimic a pond. To the west of the Dapingshan Rockery is Binzao Fengxiang Room, a half-pavilion featuring screen doors, brick benches, and a south-facing corridor leading to Xianyan Studio. Along the southern wall of Pingshan Hall grows a 130-year-old camellia tree.

凝霞阁景区
Ningxia Belvedere Scenic Area

　　凝霞阁景区是沈氏园旧址,以太湖石堆砌的大屏山为中心,北有凝霞阁,南有聊淹堂,东有扶疏堂,西有屏山堂。景区内回廊相连,院墙多置漏窗,移步易景,风雅各异。

The Ningxia Belvedere scenic area, once the Shen Family Garden, is centered around Dapingshan Rockery, an impressive formation made of Taihu rocks. Surrounding it are Ningxia Belvedere to the north, Liaoyan Hall to the south, Fushu Hall to the east, and Pingshan Hall to the west. The area is connected by winding corridors, with latticed windows along the courtyard walls that offer changing views as you stroll through, each with its own distinct charm.

丛桂轩
Conggui Gallery

舟而不游轩（池上草堂）
Zhou Er Buyou Gallery (Chishang Cottage)

行人涉桥登临桃花潭北岸，又有黄石掇山临水而起，扑水而建的碧光亭挑临潭上，秀逸温婉，亭边几株苍翠的枫杨点缀在山间水际。游人小坐于碧光亭间，倚栏西望，可见西侧的丛桂轩临水矗立，紧邻的是旱舫——舟而不游轩，同样紧依水岸，却稍稍凸前，与东岸屏山堂遥相呼应。

舟而不游轩，又名"池上草堂"，始建于清道光（1821—1850）、咸丰（1851—1861）年间，形似舟楫，却并不过度渲染旱舫的完整形态，而是着力刻画船体的凌波横流之态，船头仿若半岛，船尾隐入树丛而省略，颇具形神兼备的写意之风。池上草堂位于桃花潭西南岸，面山背水，有一形如跳板的条石连接南岸。堂中陈设明式红木家具，上悬夏雨的行书匾额。堂前有数峰玲珑石，间植桂花、海棠、芭蕉等。

Crossing the bridge to the northern bank of Taohua Pool, visitors are greeted by a rocky yellow stone formation rising beside the water, where the elegant Biguang Pavilion perches at the water's edge. The pavilion, with its delicate beauty, seems to float above the pool, framed by verdant Chinese wingnut trees. Seated within, visitors can enjoy the tranquil view, gazing westward toward Conggui Gallery, which stands by the water. Nearby is Zhou Er Buyou (a boat on the ground) Gallery, a boat-shaped structure that sits by the shore without floating, creating a picturesque connection with Pingshan Hall on the eastern bank.

Zhou Er Buyou Gallery, also known as Chishang Cottage, was built during the Qing dynasty's Daoguang (1821–1850) and Xianfeng (1851–1861) reigns. Rather than fully replicating a traditional boat, the structure captures the feeling of a boat gliding through water. The bow extends like a peninsula, while the stern blends into the surrounding trees, artfully balancing the realistic details with an abstract, poetic touch. Situated on the southwestern bank of Taohua Pool, the structure faces the hill and overlooks the water from behind. A stone slab, resembling a diving board, connects the structure to the southern bank. Inside, Ming-style *hongmu* hardwood furniture is arranged, while in front of the structure, carefully arranged rockeries are scattered amid osmanthus, crabapple, and banana trees.

桃花潭
Taohua Pool

桃花潭景区为龚氏园旧址，采用"一河两岸"式布局，以狭长的河道状水体为中心，水体一侧以少量建筑为主，另一侧以横卧的长山为主，两者隔水相对；建筑一侧以小体量轩亭突入水面，投向对岸山林，形成对狭窄水上空间的进一步分隔，凝聚起"看与被看"的焦点。景区内有池上草堂、南山、丛桂轩、碧光亭等，是秋霞圃景色的精华所在。

景区桃花潭作为中心，因旧时潭畔桃红柳绿而得名。潭水曲折潆洄，周边高林环抱，眼前寒轩亭倒映，整个世界仿佛顷刻间安静了下来。

桃花潭侧，积土掇石而成的是苍然横卧的南山。山内有平缓山谷，两侧高林之下是一条纵贯的曲径。山上有南北两岗，岗上葳蕤繁茂，北岗叠成牛、马、羊等动物形状的湖石妙趣横生。

Located on the site of the original Gong Family Garden, the Taohua Pool scenic area follows a "one river, two banks" layout, with a long, narrow waterway at its center. On one side, a few buildings line the water, while on the opposite bank, a reclining hill stretches along the shoreline. These elements face each other across the water, with small structures from the building side extending over the water toward the hill and forest, further dividing the narrow space and enhancing the interplay of "seeing and being seen." Highlights of this area include Chishang Cottage, South Hill, Conggui Gallery, and Biguang Pavilion, which showcase some of the finest scenery in Qiuxia Garden.

Taohua Pool, the focal point of this area, earned its name from the peach blossoms and willow trees that once adorned its banks. The meandering waters are framed by tall trees, and the reflection of Hanxuan Pavilion in the pond creates a serene scene, evoking a peaceful silence as if time has momentarily paused.

On the side of Taohua Pool, South Hill sprawls across the landscape, formed from piled earth and rocks, giving it a rugged, natural appearance. A gentle valley runs through the hill, flanked by towering trees, with a winding path weaving through the greenery. The hill features two ridges — north and south — both lush and vibrant. The northern ridge is particularly charming, with lake rocks arranged to resemble animals such as cows, horses, and sheep, adding a whimsical touch to the landscape.

桃花潭景区
Taohua Pool Scenic Area

上：碧光亭
U: Biguang Pavilion
下：南山
D: South Hill

园景撷趣

Garden Scenery

扶疏堂
Fushu Hall

"现代中国园林之父"陈从周先生曾在其《嘉定秋霞圃和海宁安澜园》一文中详细介绍过秋霞圃，并于 20 世纪 80 年代在古园百废待兴之时，指导了秋霞圃的修复工作，足见秋霞圃在园林专家心中的重要地位。当我们寻踪园林哲匠造园的精妙用意时，总会感慨秋霞圃"一庙三园"格局的优美和明风古园"一河两岸"美景的深邃。

Chen Congzhou, known as the "Father of Modern Chinese Gardens," devoted significant attention to Qiuxia Garden in his work *Qiuxia Garden in Jiading and Anlan Garden in Haining*. In the 1980s, when the garden required extensive restoration, Chen personally oversaw the project, highlighting its esteemed status among garden experts. Visitors can appreciate the elegance of the "one temple, three gardens" layout and the tranquil beauty of the "one river, two banks" landscape — hallmarks of the ancient Ming garden tradition.

part of the rear grounds of the City God Temple. The Jin Family Garden, located north of the Gong Family Garden, was developed by three generations of the Jin family — Jin Yi, Jin Dayou (*Juren*, a successful candidate of the provincial imperial examinations), and Jin Zhaodeng — during the Jiajing period (1522–1566) of the Ming dynasty.

In 1860, during the Taiping Heavenly Kingdom Movement, Qiuxia Garden suffered severe damage, with many of its buildings destroyed. Reconstruction began in 1876 during the Guangxu reign (1875–1908) and continued over the following decades. In 1920, Headmaster Dai Sigong of Qiliang School relocated the school from inside the East Gate to the eastern section of the City God Temple's rear garden, which had been the site of the Shen Family Garden. Dai also enlisted local business leaders to sponsor the rebuilding of the gardens' structures. By 1922, more than 20 structures, including Liaoyan Hall, Youchi Hall, Fushu Hall, and Jishan Pavilion, had been restored. In 1937, the Japanese occupation led to the closure of Qiliang School, and the garden was taken over by occupying forces. After Japan's surrender in 1945, the garden was briefly occupied by the Three Principles of the People Youth League Jiading branch and later became the office of Jiading County's First District. In May 1946, the district office relocated, and by June, the garden was renamed "Yimiao Park" and opened to the public. Qiliang School reopened in August, and management of the garden's buildings was handed over to the school. Over the years, Qiuxia Garden experienced numerous changes, including significant damage during the Cultural Revolution. Beginning in 1980, the municipal and county governments invested over 3 million yuan in a two-phase restoration, rebuilding and enhancing 48 scenic spots. On October 1, 1987, the restoration was completed, and the garden was officially reopened to the public as "Qiuxia Garden."

Today, Qiuxia Garden is composed of four distinct scenic areas: the Gong Family Garden (Taohua Pool scenic area) covering approximately 8 *mu* (approximately 5333 square meters); the Shen Family Garden to the east (Ningxia Belvedere scenic area) covering approximately 4 *mu* (approximately 2667 square meters); the Jin Family Garden to the north (Qingjing Pond scenic area) covering approximately 20 *mu* (approximately 13 333 square meters); and the City God Temple scenic area in the southeast. Altogether, these areas span a total of 45.36 *mu* (30 240 square meters). In February 2023, Qiuxia Garden was awarded a five-star park rating, reflecting its historical significance and beauty.

归云洞
Guiyun Cave

Located at No. 314 East Street in Jiading Old Town, Qiuxia Garden is admired for its intricate layout and tranquil surroundings, earning it the title of an "oasis in the heart of the city."

Qiuxia Garden traces its origins to the early 16th century, during the Ming dynasty's Zhengde (1506–1521) and Jiajing (1522–1566) periods. It was first established by Gong Hong (1451–1526), a high-ranking official who served as Grand Master for Glorious and minister of the Ministry of Works. Originally known as the "Gong Family Garden," it was located behind Gong's residence. In 1555, the 34th year of Jiajing period, after Gong's great-grandson, Gong Minqing, was tragically killed by a servant in the bondservant uprisings, the garden briefly passed into the hands of Wang, a merchant from Huizhou. However, in 1573, following Gong Minqing's son, Gong Xijue's success in the imperial examinations, Wang honored his promise and returned the garden to the Gong family.

During the tumultuous Ming-Qing transition, the Gong family faced great losses, with many male members sacrificing their lives in resistance to the Qing. The devastated family lost the estate again, and during the Shunzhi (1644–1661) and Kangxi (1662–1722) periods, it was reclaimed by the Wang family. The new owner, Wang Yuwu, expanded the garden by adding buildings, plants, and rockeries, renaming it "Qiuxia Garden," also known locally as the "Wang Family Garden." The name "Qiuxia" (autumn clouds) was inspired by a famous line from *Preface to the Prince of Teng's Pavilion*, written by Wang Bo of the Tang dynasty: "Rosy clouds and a lonely wild duck fly together; the autumn waters merge with the vast sky into one hue," reflecting the owner's unyielding ambitions. During this period, Qiuxia Garden became a cultural haven where scholars and literati from Jiangnan and beyond gathered to exchange ideas, write poetry, and paint, making it a vibrant center of cultural activity.

In 1726, during the Qing dynasty's Yongzheng period (1723–1735), the Wang family, having fallen into decline, donated the Qiuxia Garden, which then became part of the rear garden of the City God Temple. The City God Temple was originally built during the Jiading period (1218–1224) of the Southern Song dynasty in the southern part of the city. In 1370, during the Ming dynasty's Hongwu reign, it was relocated to its current site in the eastern part of the city. Over the next 500 years, the temple underwent multiple reconstructions, with major expansions in 1726 and 1755. These expansions incorporated both the Wang family's Qiuxia Garden to the west and the Shen Family Garden to the north, transforming them into part of the temple's Sacred Garden. During the Republic of China period, the City God Temple was repurposed as a military barracks, resulting in the destruction of nearly all its statues and buildings. The site was later taken over by Qiliang School and, starting in the 1950s, served as a school building for many years.

Today, Qiuxia Garden includes two additional sections: the Shen Family Garden and the Jin Family Garden. The Shen Family Garden was originally established during the Ming dynasty's Tianqi period (1621–1627) by Shen Hongzheng (1578–1627), a scholar who purchased a neglected part of the Gong Family Garden to the east. It later passed into the hands of the Shen family. During the Qing dynasty's Qianlong reign (1736–1795), the Shen Family Garden was incorporated into Qiuxia Garden, becoming

署迁出，6月定名为"邑庙公园"对外开放。8月启良学校复校，县政府将园内所有建筑拨给启良学校管理使用。此后，秋霞圃又几经更迭，尤其在"文革"中再度严重损毁。1980年起，市、县政府耗资300余万元，统一规划，分两期重修、重建、新建景物48处。1987年10月1日，全面竣工开放，统称"秋霞圃"。

如今的秋霞圃，由三园一庙共四个景区组成，包含龚氏园（桃花潭景区）约8亩（约5333平方米）、东侧沈氏园（凝霞阁景区）约4亩（约2667平方米）、北侧金氏园（清镜塘景区）约20亩（约13 333平方米），以及东南侧城隍庙景区，现占地总面积为45.36亩（30 240平方米）。2023年2月，秋霞圃被评定为上海市五星级公园。

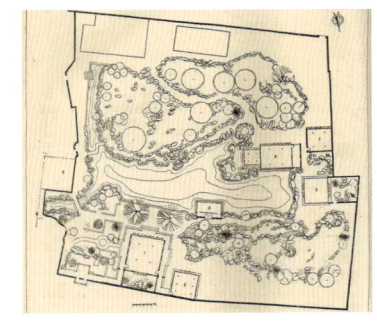

秋霞圃平面图，20世纪60年代初（陈从周绘，上海市嘉定博物馆提供）
Plan of Qiuxia Garden, early 1960s (Painted by Chen Congzhou, provided by Shanghai Jiading Museum)

秋霞圃一隅
A corner of Qiuxia Garden

以文会友、吟诗挥毫，繁盛一时。

清雍正四年（1726），汪氏衰弱，遂把秋霞圃捐出，改属城隍庙后园。此处城隍庙又称"城隍祠"或"邑庙"，初建于南宋嘉定年间（1208—1224），原坐落于南城富安坊中，明洪武三年（1370）移建至东城今址。至清光绪八年（1882）的500多年间，邑庙多次重修重建，特别是在1726年和1755年前后，有过两次大扩容，即当时庙西汪氏秋霞圃与庙北沈氏园先后并入，成为邑庙灵苑。民国时期，城隍庙成为驻军之处，殿内神像及殿外建筑十之八九被毁，后归启良学校，20世纪50年代起被长期用作校舍。

如今秋霞圃的另外两处组成部分为沈氏园和金氏园。沈氏园是明天启年间（1621—1627），诸生沈弘正（1578—1627）购得龚氏园东面一处偏废园圃构建而成，后归申氏所有。清乾隆年间（1736—1795），沈氏园并入秋霞圃，同为城隍庙的后园。金氏园位于龚氏园之北，为明嘉靖年间

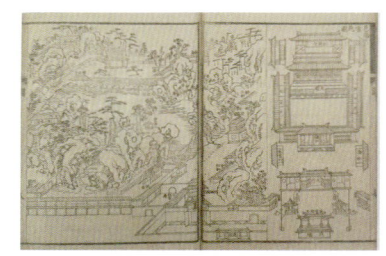

清嘉庆《嘉定县志》邑庙灵苑图（上海市嘉定博物馆提供）
Sacred Garden Map, Gazetteer of Jiading County, the Jiaqing Period of the Qing dynasty (Provided by Shanghai Jiading Museum)

（1522—1566）金翊、举人金大有、金兆登祖孙三代营造。

清咸丰十年（1860）因太平天国兵燹，秋霞圃受劫，亭台楼阁毁于一旦。光绪二年（1876）后，秋霞圃陆续重建。1920年邑人戴思恭校长将原设在东门内的启良学校迁入邑庙后园东部，即沈氏园故址，并发动城区实业界分别认建、认修园内亭台楼阁。至1922年先后修建了聊淹堂、游骋堂、扶疏堂、即山亭等20余座建筑。1937年日军入侵，邑庙后园被敌伪所占，启良学校停办。1945年日军投降后，邑庙后园被三青团嘉定分团部占用，2个月后迁出，设为嘉定县第一区区署。1946年5月，第一区

历史沿革

Garden History

秋霞圃位于嘉定老城东大街314号,园内布局精致,环境幽雅,被誉为"城市中的山林"。

明正德(1506—1521)、嘉靖年间(1522—1566),荣禄大夫工部尚书龚弘(1451—1526)创园于宅后,当时称"龚氏园"。嘉靖三十四年(1555),龚弘曾孙龚敏卿被家奴杀害后,宅园曾短暂归徽商汪氏所有。万历元年(1573),龚敏卿之子龚锡爵中举后,汪氏履行承诺,将此园归还于龚氏。

明清更迭之际,龚氏男丁为抗清牺牲惨烈、家毁人亡,清顺治(1644—1661)、康熙(1662—1722)年间,汪氏再次接手龚氏宅园。当时主人汪于梧将其扩建为园林,添建木石亭馆,易名为"秋霞圃",俗称"汪氏园"。"秋霞"取自唐代王勃名篇《滕王阁序》中的"落霞与孤鹜齐飞,秋水共长天一色",寄托着原园主"不坠青云之志"的愿望。当时江南及各地文人雅士常聚于此,

明龚弘像(清代程祖庆《练川名人画像》,上海市嘉定博物馆提供)
Portrait of Gong Hong from the Ming dynasty (*Portraits of Famous People from Jiading Lianchuan*, Cheng Zuqing, Qing dynasty, provided by Shanghai Jiading Museum)

20世纪60年代初的秋霞圃(陈从周摄,上海市嘉定博物馆提供)
Qiuxia Garden, early 1960s (Photo by Chen Congzhou, provided by Shanghai Jiading Museum)

柳云居
Liu Yun Residence

秋霞圃始建于明代,是上海著名的五大古典园林之一,由龚氏园、沈氏园、金氏园与城隍庙后园(含城隍庙)合并而成。园内含 48 处名胜古迹,有上海"最古老园林"之名。

First constructed during the Ming dynasty, Qiuxia Garden is one of Shanghai's five renowned classical gardens. It was formed by combining the Gong, Shen, and Jin family gardens, along with the rear garden of the City God Temple. With 48 scenic spots, it holds the distinction of being the oldest garden in Shanghai.

秋霞圃
Qiuxia Garden

栉风沐雨，五百余载——秋霞圃

Over Five Centuries of Resilience: Qiuxia Garden

年代：明代
地址：上海市嘉定区东大街314号
占地面积：30 240平方米
保护级别：上海市文物保护单位

Era: Ming dynasty
Address: No. 314, East Street, Jiading District, Shanghai
Area: 30 240 square meters
Protection level: Shanghai city-level protected site

秋霞圃鸟瞰图
Aerial View of Qiuxia Garden

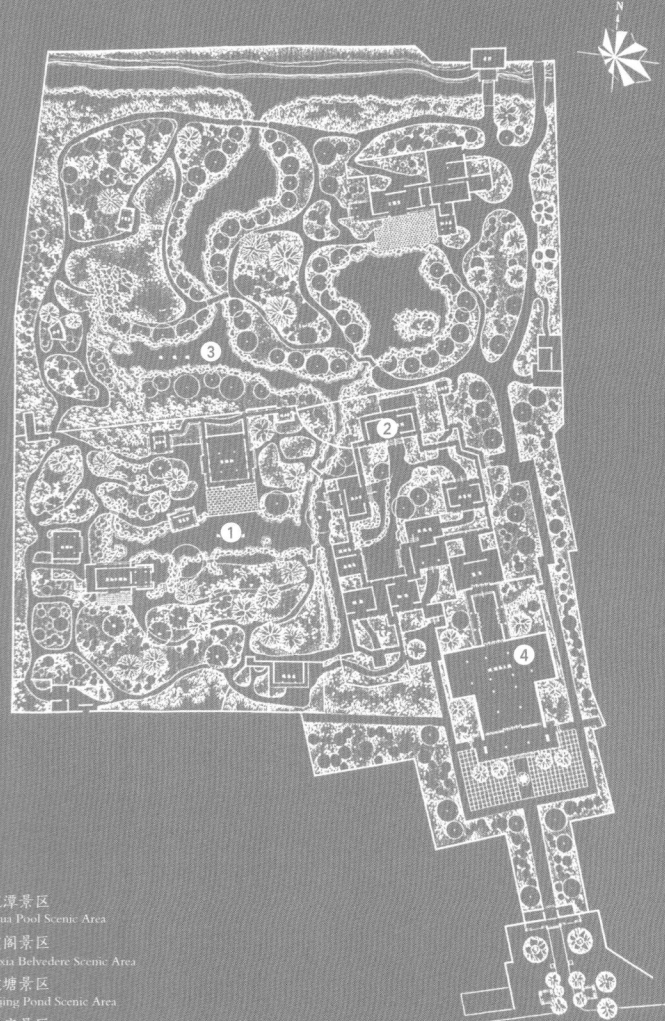

图例
Legends

1 桃花潭景区
 Taohua Pool Scenic Area

2 凝霞阁景区
 Ningxia Belvedere Scenic Area

3 清镜塘景区
 Qingjing Pond Scenic Area

4 城隍庙景区
 City God Temple Scenic Area

秋霞圃

Qiuxia Garden

豫园南门
South gate of Yu Garden

Changshuo, and Pu Zuoying — founded the Yu Garden Calligraphy and Painting Association at Deyue Tower, where they curated and exhibited art collections regularly.

Unfortunately, the 1937 Battle of Shanghai inflicted further damage on Yu Garden. It wasn't until 1956, after 1949 and the onset of a stable society, that the first comprehensive restoration of the garden took place. In the 1980s, with the advancement of cultural heritage preservation, a second major restoration was undertaken. This effort focused on balancing historical preservation with modern utility, leading to the reconstruction of the eastern section and the introduction of new landscapes. These enhancements not only safeguarded the garden as a cultural heritage site but also elevated its status as a tourist destination and cultural symbol. This philosophy continues to guide Yu Garden's evolution, ensuring that it remains not only an enduring legacy but also a vibrant and dynamic emblem of the city.

From its beginnings as a private garden in the early Ming dynasty, through its transformation into a guild garden in the late Qing dynasty, to its current role as a public garden, Yu Garden has continually evolved in its land use, functions, layout, and name. This ongoing process of destruction and renewal has layered historical elements into the garden's landscape, reflecting its spirit of inclusiveness, innovation, and adaptability. Situated in the heart of Shanghai, Yu Garden has witnessed the city's transformation from an agricultural society to a commercial economy, endured the turmoil of the late Qing and Republic of China periods, and adapted to the societal changes of modern China. Today, this historic garden thrives in the bustling city center, offering an enchanting oasis that captures the essence of Jiangnan's landscape within an urban setting. As a cherished city landmark, Yu Garden will continue to enhance Shanghai's character, adding beauty and color to the city's evolving narrative.

种理念也贯穿至今，让豫园不仅是"凝固的历史"，也是充满生机的城市名片。

从明代初建时期的私家园林，到晚清时期的公所园林，最终演变为市民园林，在此过程中，豫园的用地、功能、格局、名称一直变动，不同时期的历史信息伴随毁弃和添置的过程层叠增添至园林场景中[4]，表现出了兼容并包、求新求变的特征。处于上海城市中心地区的豫园，见证了上海从农业社会到商品经济社会的转变，亲历了清末至民国时期的战火纷飞，也在新中国的社会建设中体味着时代变迁，如今，这座海上名园身处闹市，创造着城市山林的美妙图景，呈现着繁华都市中的江南风光。它也将继续作为城市地标，为城市面貌增色，为城市画卷添彩。

---

1 同济大学建筑与城市规划学院景观学系. 陈从周造园三章[M]. 上海: 同济大学出版社, 2018: 4.
2 朱宇晖. 江南名园指南（上）[M]. 上海: 上海科学技术出版社, 2002.
3 陈业伟. 豫园 [M]. 上海: 上海文化出版社, 2009: 184.
4 段建强. 陈从周先生与豫园修复研究 [C]// 中国建筑学会建筑史学分会, 华南理工大学建筑学院.《营造》第五辑——第五届中国建筑史学国际研讨会会议论文集（下）. 2010: 8.

Yu Garden, located in the heart of the city, has a history deeply intertwined with the tides of time. Originally built during the prosperity of the Pan family, the garden faced decline following the family's fall and the social upheaval of the late Ming dynasty. During the Qianlong period (1736–1795), it was entrusted to the Daoists of the City God Temple and gradually transitioned from a private retreat into a public garden. Despite suffering damage during the social unrest and wars of the Jiaqing and Daoguang periods (1796–1850), these events also brought about significant changes, transforming Yu Garden into a venue for political, diplomatic, and commercial activities. It hosted guild meetings, gentry gatherings, proclamations of imperial edicts, celebrations of the emperor's birthday by local officials, and banquets for both domestic and international guests. Yu Garden also emerged as one of Shanghai's cultural and artistic hubs. Renowned painters such as Ren Bonian, Xugu, and Wu Changshuo left their mark here. In 1909, during the first year of the Xuantong reign, prominent figures in Shanghai's calligraphy and painting community — Gao Yong, Qian Hui'an, Wu

任伯年为豫园点春堂所作的《观剑图》（上海市历史博物馆提供）
Observing the Sword (Ren Bonian, provided by Shanghai History Museum)

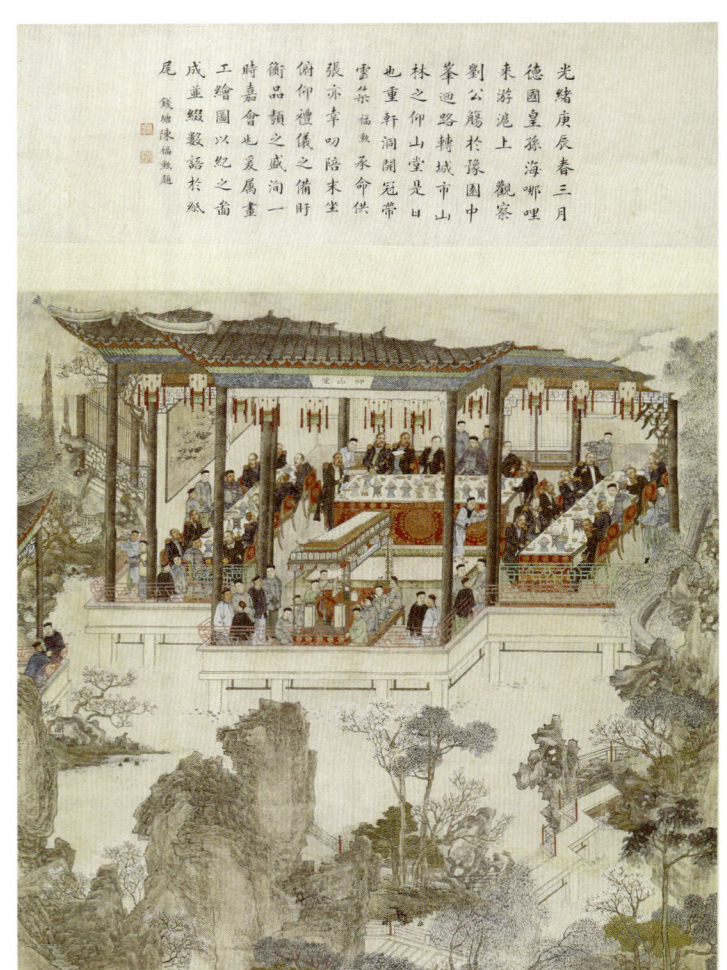

光绪庚辰春三月
德国皇孙海哪哩
来游沪上观察
刘公篤于豫园中
峯迴路转城市山
林之仰山堂是日
也重轩洞開冠带
雲集馔魚承命供
张亦幸叨陪末坐
俯仰禮儀之備时
衡品题之盛洵一
時嘉會也爰属一
工繪圖以紀之畬
成並缀數語於紙
尾　錢塘陳福熏題

吴友如《豫园宴乐图》（上海市历史博物馆提供）
Festive Banquet at Yu Garden (Wu Youru, provided by Shanghai History Museum)

人文荟萃
Gathering of Talents

豫园位于城市中心，与时代的命运紧密相连。它因潘家的兴旺而诞生，又因明末社会动荡、潘家的没落而衰退。自清乾隆年间（1736—1795）托管给城隍庙道士后，它逐步从私人园林转变为公共园林。清嘉庆（1796—1820）、道光年间（1821—1850），社会局势动荡，战争给豫园带来了创伤，但也给豫园带来了新的转变。行业议事、士绅聚会，乃至宣讲"圣谕"、朝贺"万寿"（道县两级官员遥贺皇帝诞辰），以及宴请国内外宾客等市政公共事务均在此发生，使豫园成为政治、外交、商贸的舞台。同时，豫园还成为沪上的文化艺术中心之一。任伯年、虚谷、吴昌硕等画坛巨擘先后在这里泼墨敷彩。清宣统元年（1909），海上书画名家高邕、钱慧安、吴昌硕、蒲作英等人在豫园得月楼发起豫园书画善会，定期收集画作举办展览。

可惜，淞沪会战再一次造成了园景的破坏。直到社会稳定发展的 1956 年，才开始了第一次整体性修缮。20 世纪 80 年代，文物保护理念的成熟又促成了第二次大修。出于对历史保护的态度、兼顾豫园在当代的利用，豫园重建了东部片区，塑造了新景观，使其作为一处文物得到更好的保护，作为一处旅游胜地和文化符号得到充分的展现。这

参观指南 Visiting Guide

开放时间： 09:00—16:30（16:00 停止入园，16:30 闭园）；每周一闭园（国家法定节假日除外，春节期间开放时间以公告时间为准）。
交通信息： 地铁 10 号线、14 号线至豫园站；公交 11 路、26 路、64 路、736 路、926 路、929 路、932 路、969 路、980 路、方川线等。

Opening hours: 9:00 – 16:30 (admission stops at 16:00 and closes at 16:30); The park is closed every Monday (excluding national holidays, and the opening hours during the Spring Festival are subject to the announced time).
Traffic information: Metro Line 10 and Line 14 to Yuyuan Garden Station; Bus 11, 26, 64, 736, 926, 929, 932, 969, 980, Fangchuan Line, etc.

和煦堂内的清代榕树根家具
Qing-dynasty banyan root furniture in Hexu Hall

铁狮子
Iron lion

intricate figures and layered details, showcase the masterful craftsmanship and artistic vision that bring the garden's decorations to life.

The fifth defining feature of Yu Garden is the thoughtfully placed plants and trees that add the perfect finishing touches, bringing the garden to life. On the east side of Sansui Hall, an ancient Luohan pine stands prominently at the entrance of the Jianru Jiajing Corridor, drawing the eye and guiding visitors' views. In the Inner Garden, a Chinese hackberry tree fills the narrow gap between the Dongtian Fudi scenic spot and the Keyi Hall, seamlessly bridging the space between the two buildings. In front of Yuhua Hall, two white magnolia trees bloom with pure white flowers, their fragrance filling the air. The ancient ginkgo tree in front of Wanhua Tower, with its sprawling roots and expansive canopy, is regarded as the treasure of the garden. These carefully chosen plants and trees not only enhance the sense of discovery along the garden's winding paths but also contribute to the overall beauty and charm of the landscape, making each corner of the garden a delight to explore.

The wonders of Yu Garden extend beyond these five features. Scattered throughout the garden, eight pairs of lions sit before various halls and gateways, each with a unique and playful posture. The Ming-dynasty *zitan* hardwood furniture in Yuhua Hall and the Qing-dynasty banyan root furniture in Hexu Hall are exquisite treasures that embody understated elegance. These elements, along with the garden's architecture, waterscapes, and plantings, collectively contribute to the craftsmanship and charm that define this renowned garden by the sea.

拱式水花墙
An arched wall

"穿云龙"泥塑
"Dragon among Clouds" Clay Sculpture

"二龙戏珠"泥塑
"Double Dragons Playing with a Pearl" Clay Sculpture

Corridor, shaped like a beauty's waist; and Wulaofeng in front of Deyue Tower, which resembles five elders. These rocks, with their varied forms and locations, complement and enhance each other, playing a vital role in defining spaces and shaping the landscape throughout the garden.

The third defining feature of Yu Garden is its artful use of water, which flows and converges throughout the landscape, creating a seamless connection across the garden. North of Sansui Hall, an open pond splits in two directions — one stream flows north toward the Grand Rockery, while the other flows east toward Wanhua Tower, gently passing through an arched wall to reach Dianchun Hall. In the Huijing Tower area, the water expands again, linking three sections with two bridges, creating a layered sequence of spaces complemented by Huijing Tower, Liushang Pavilion, and Jiushi Gallery. As the water reaches the Yuhua Hall area, it is divided by the Jiyu Waterside Corridor, adding variation and movement before it finally flows into Jiulong Pond in the Inner Garden. These interconnected water bodies are thoughtfully arranged to follow the natural terrain, linking the garden's various landscapes, enhancing its visual appeal, and adding a lively dynamism to the space.

The fourth defining feature of Yu Garden is its intricate and lifelike decorations, which enliven the space. Chief among these are the clay sculptures on the surrounding walls, featuring motifs such as "dragons among clouds" "crouching dragons" "sleeping dragons," and "double dragons playing with a pearl." The dragon heads rise above the walls, with their bodies formed from roof tiles, creating the illusion of dragons soaring through the sky. These dynamic sculptures add movement and an auspicious atmosphere to the garden's serene landscape. In addition to the dragon walls, the garden is adorned with exquisite brick carvings. Notable examples include depictions of the Eight Immortals and the Willow Spirit along the long corridor east of Huijing Tower, as well as the Palace of the Moon on the eastern corridor wall of Qizao Hall. These carvings, with their

侧渐入佳境游廊口的罗汉松屹立在要道中间，引领视线；内园洞天福地与可以观之间缝隙中的朴树，填补了两栋建筑之间狭小间距的空白；玉华堂前对植的两棵白玉兰树花开时满枝白色，沁人心脾。万花楼前的古银杏盘根错节，绿荫蔽天，堪为镇园之宝[3]。这些花木的搭配使人在曲径通幽处更觉"柳暗花明"，也使园林的景观更为优美动人。

自然，豫园的精彩之处不仅有这五个方面。园内八对狮子坐卧几处厅堂和门洞前，姿态各异、俏皮可爱。玉华堂内的明代紫檀家具、和煦堂内的清代榕树根家具等珍品，尽显低调中的奢华。它们与园内建筑、水体、花木共同构成了这座海上名园的匠心韵味。

The brilliance of Yu Garden lies in five distinctive features.

The first distinctive feature of Yu Garden is its masterful use of space, creating "a world within a small space." Every inch of the garden is thoughtfully designed, with buildings, rock formations, plants, and waterscapes carefully arranged to form spaces of varying scales that interact harmoniously. For example, the open, expansive view in front of the Grand Rockery contrasts strikingly with the intimate, secluded space behind it at Cuixiu Hall. The imposing presence of Wanhua Tower is softened by the small, delicate Liangyi Gallery in front, providing a smooth transition in scale. Throughout the garden, dragon walls with undulating gray tiled ridges define the boundaries of different areas, transforming each small space into a self-contained scene rich in layers. Latticed windows, rockeries, and other garden elements serve as transitional markers, enhancing the fluidity and dynamism of the spaces, adding to the garden's overall charm.

The second defining feature of Yu Garden is the rugged yet graceful beauty of its rocks, which play a central role both in the piled-up rock hills and as singular, uniquely shaped rocks. Beyond the Grand Rockery in front of Yangshan Hall, other rockeries are strategically placed throughout the garden. For example, Baoyun Rockery in front of Kuai Tower is vertically stacked, resembling a peak rising sharply from the water, mirroring the tower's tall and elegant stance. The Huanyun Rockery near Huijing Tower includes winding caves that lead to the summit, with a spring flowing through, creating a scene of natural splendor. In addition to the renowned Yulinglong, other notable rocks include Jiyufeng in the Jiyu Waterside Corridor, which resembles a mound of jade; Meirenyao in the Jianru Jiajing

九狮轩前水体
Waterscape in front of Jiushi Gallery

"美人腰"石
Meirenyao Rock

谷音洞
Guyin Stream

三是水之有聚有散。豫园内的所有水体都是连贯的。三穗堂以北的开阔水池分流两个方向,向北流向大假山,向东流入万花楼,再悠悠荡荡穿过一道拱式水花墙至点春堂。到会景楼片区,水体面积再次增大,以两座桥连接三片水域,会景楼、流觞亭、九狮轩三种不同类型的建筑穿插其中,形成连续且层次丰富的空间序列。至玉华堂景区内,水体又以积玉水廊进行分割,增加变化和动感,最终流向内园九龙池。这些水体循地形而安排,串联起了整个园林的景观,丰富了环境色彩,增加了整个园子的生气。

四是装饰之精妙传神。如围墙上装饰的"穿云龙""卧龙""眠龙""双龙戏珠"泥塑。五龙头高于墙,以脊瓦为身,仿佛在空中游动,为静态的园景增加了动感和吉祥瑞福的氛围。除了龙墙外,园内还有很多精美的砖刻,如会景楼东侧长廊上的"八仙"与柳树精、绮藻堂东廊壁间的"广寒宫",人物众多,层次分明,匠心独特。

五是花木之画龙点睛。三穗堂东

快楼与抱云岩
Kuai Tower and Baoyun Rockery

龙墙
The Dragon Walls

价值特色

Features and Significance

纵观豫园精妙之处，可归纳为五个方面。

一是空间之咫尺乾坤。豫园的每一寸土地都得到了充分利用，以建筑、山石、花木、池塘等围合形成不同尺度的空间，碰撞出趣味。例如，大假山前的开敞山水与大假山后萃秀堂幽静的狭小空间形成鲜明对比；大体量的万花楼前布置小巧的两宜轩作为体量上的过渡；园内的多处龙墙界定了各区的边界，使各个小空间独立成景，层次丰富，同时利用漏窗、山石等园林小品作为空间理景的转换符号，增加了空间的动感。

二是石之嶙峋俊美。豫园中的石主要见于堆叠而成的假山与独立摆放的湖石。假山除仰山堂前的大假山外还有多处，例如，快楼前的抱云岩，竖向堆叠，似山峰从池水中拔地而起，与快楼挺拔之姿相呼应。会景楼片区的浣云假山内山洞盘旋，循洞可达山顶，洞内有泉，涧溪水从洞内流出，颇具大自然之美。湖石除了声名卓著的"玉玲珑"外，另有"积玉峰"立于积玉水廊内，宛若美玉堆砌；"美人腰"在"渐入佳境"游廊内，形似婀娜佳人；"五老峰"在得月楼前，状如耄耋老人。这些不同地点、不同姿态的石头互相衬托呼应，在空间分隔、景观塑造方面发挥了重要作用。

复廊
Double-Lane Corridor

玉玲珑再往南可至最后一个景区——内园，内园即原城隍庙庙园，为清康熙四十八年（1709）所建，占地仅2亩（约1333平方米），这里的园林风格和前述的片区不同，踏入院墙，便能感受到其在更为紧凑空间中的错落有致。园内以"静观"厅为中心建筑，假山为景观引导。园中有观涛楼，高三层，相传登楼可观黄浦江。园内九龙池面积虽小，但从龙墙下洞曲流出，意境深远。另有"可以观""还云楼""延清楼"等，颇具"螺蛳壳里做道场"的趣味。

South of Yulinglong lies the southernmost section of Yu Garden — the Inner Garden. Originally part of the City God Temple's garden, the Inner Garden was established in 1709 during the Kangxi period of the Qing dynasty. Although it covers only 2 *mu* (approximately 1333 square meters), its design contrasts with the more expansive areas of Yu Garden, showcasing a thoughtful and efficient use of compact space. At the heart of the Inner Garden is Jingguan Hall, with a rockery guiding the overall landscape. One of its key features is Guantao Tower, a three-story building from which, according to legend, one could once view the Huangpu River. Despite its small size, Jiulong Pond adds a sense of depth, with waters flowing from beneath dragon walls through winding channels, creating a tranquil atmosphere. Other notable structures, such as Keyi Hall, Huanyun Tower, and Yanqing Tower, each contribute to the garden's unique charm and exemplify the art of crafting a grand experience within a limited space.

观涛楼
Guantao Tower

内园门楼
Gate of Inner Garden

内园
Inner Garden

上：静观
U: Jingguan Hall
下：还云楼
D: Huanyun Tower

和煦堂西侧是会景楼。会景楼建于20世纪60年代，原名"得意楼"。经会景楼向南过一小桥可至玉华堂和豫园最著名的奇石——玉玲珑。玉玲珑相传是宋徽宗时期花石纲流散之物。峰石玲珑剔透，孔多如蜂巢，可呈现"百孔淌泉，百孔冒烟"的奇观，具有漏、皱、瘦、透之美，是江南三大名石之一。此石原在浦东三林塘太仆寺少卿储昱之南园，储昱之女嫁潘允端之弟潘允亮后携此石至潘家。水运途中船翻石沉，后在打捞过程中又发现一石可配为玉玲珑底座，便共同放置在豫园中。清代陈维城《玉玲珑石歌》中有云："一卷奇石何玲珑，五丁巧力夺天工。不见嵌空皱瘦透，中涵玉气如白虹。[2]"

West of Hexu Hall stands Huijing Tower, a structure built in the 1960s, originally named Deyi Tower. Crossing a small bridge to the south, you arrive at Yuhua Hall and one of Yu Garden's most famous features — Yulinglong. According to legend, Yulinglong dates back to the rock collection efforts of Emperor Huizong during the Song dynasty, a project to gather uniquely shaped rocks from coastal provinces. The rock is renowned for its intricate design, with numerous beehive-like holes capable of producing the stunning effects of "a hundred holes dripping water, a hundred holes emitting smoke." It exemplifies the ideal qualities of being leaky, crumpled, slender, and holey, making it one of the three most famous rocks in the Jiangnan region. Originally, Yulinglong was in the South Garden of Chu Yu, a Vice Minister of the Court of the Imperial Stud, in Sanlintang, Pudong area. After Chu Yu's daughter married Pan Yunliang, the brother of Pan Yunduan, the rock was brought to the Pan family. During its transportation by water, the boat carrying the rock capsized, causing it to sink. During the salvage, another stone was discovered to serve as its base, and the two were placed together in Yu Garden. The Qing dynasty poet Chen Weicheng praised the Yulinglong in the poem *Song of the Yulinglong Rock*, saying: "How exquisite, this coiled and wondrous rock! Rivaling nature's craft with artisans' skill alone. Its holey surfaces, crumpled texture, slender grace. Revealing inner beauty, like jade's radiant trace."

会景楼与玉玲珑
Huijing Tower and Yulinglong

上：会景楼
U: Huijing Tower
下：玉玲珑
D: Yulinglong

万花楼东侧，便是点春堂。点春堂建于清道光初年，得名于苏轼词中的"翠点春妍"。该堂为五开间、单檐歇山顶建筑。室内雕梁画栋，装饰华丽。点春堂北面临水，有一座凸出的水阁，可于其上观池鱼游跃，南侧有一座方形厅堂——和煦堂，该堂四面敞开，春日阳光可尽洒室内，给人舒畅的体感，遂以"和煦"命名。

Continuing east from Wanhua Tower, you'll arrive at Dianchun Hall. Built in the early Daoguang period of the Qing dynasty, the hall's name is inspired by a line from Su Shi's poem, which describes the fresh greenery and blossoms of the spring. The hall is a five-bay structure with a single-eaved hipped roof, featuring intricately carved beams and lavish interior decorations. North of Dianchun Hall is a waterside belvedere that extends over a pond, offering a serene spot to observe the fish swimming below. To the south lies Hexu Hall, a square building open on all four sides. In spring, sunlight fills the hall, creating a warm and welcoming atmosphere, which inspired its name, "Hexu," meaning "warm and gentle."

点春堂内景
Interior of Dianchun Hall

和煦堂与点春堂内景
Interior of Hexu Hall and Dianchun Hall

点春堂与和煦堂
Dianchun Hall and Hexu Hall

上：点春堂
U: Dianchun Hall
下：和煦堂
D: Hexu Hall

大假山以东是万花楼景区。进入万花楼前，首先会经过一道复廊。廊中用墙分隔为左右两条通路，墙上开设砖框与漏窗，可移步换景，左顾亦舫，右看两宜轩，享受丰富的视觉体验。复廊尽头可见万花楼。万花楼所在位置为明代"花神阁"遗址，清代改建西园后，此处称为"万花深处"，因供奉过城隍老爷，又名"神尺堂"，寓意"人神仅咫尺相隔"。现存主体建筑是清道光年间（1821—1850）重建，高两层，精雕细镂，造型美观。楼前一株古银杏已有 400 多年历史，枝叶繁茂，相传为潘允端父亲潘恩手植，是豫园中最为古老的树木。

East of the Grand Rockery lies the Wanhua Tower scenic area. Before reaching the tower, you pass through a double-lane corridor, divided by a central wall into two parallel walkways. The corridor features brick-framed latticed windows that offer varying views as you move through. To the left, you can glimpse Yi Boat Hall; to the right, Liangyi Gallery, providing a rich and immersive visual experience. At the end of the corridor, Wanhua Tower comes into view. The tower stands on the site of Flower God Belvedere of the Ming dynasty. After West Garden's renovation during the Qing dynasty, this area became known as *"Wanhua Shenchu"* (literally "Deep in the Flowers") and later as Shenchi Hall when it housed an altar to the City God, symbolizing the close connection between humans and deities. The current structure, rebuilt during the Daoguang period (1821–1850), is a two-story building adorned with intricate carvings and elegant design. In front of Wanhua Tower stands an ancient ginkgo tree, believed to be over 400 years old and planted by Pan Yunduan's father, Pan En, making it the oldest tree in Yu Garden.

万花楼
Wanhua Tower

上：万花楼
U: Wanhua Tower
下：万花楼内景
D: Interior of Wanhua Tower

从豫园南门进入，迎面可见三穗堂。"三穗"出典于"梁上三穗"，寓意官运亨通，也蕴含对丰收的希冀和祝福。建筑高9米，屋顶为单檐歇山式，与一般建筑不同的是，屋脊上的两个塑像不是鱼形鸱吻，而是拿长矛的张飞和持大刀的严颜。三穗堂室内高悬"三穗堂""灵台经始""城市山林"三块贴金匾额，讲述该堂的得名缘由和园主的审美旨趣。

三穗堂以北有大假山，大假山是豫园精华之所在，为江南现存明代黄石假山之最巨者，高约14米，"峻嶒秀润，颇惬观赏"，出自名匠师张南阳之手。张南阳以画家而业叠山，所构此山多丘壑之美。可惜现已不让登临，只能远观遐想置身山岭之趣。假山之上有一座凉亭"望江亭"，坐此四望可得全园之景。以今人之眼光虽算不得多高，但早时却是上海人登高远眺黄浦江之地，每年农历九月初九皆游人如织。

As you enter Yu Garden through the South Gate, Sansui Hall stands directly ahead. The hall's name, derived from the story "Three Ears of Grain on the Beam," symbolizes both the hope for a successful career and wishes for a bountiful harvest. The hall rises 9 meters tall, with a single-eaved hipped roof. Unusually, the two sculptures on its roof ridge are not the typical fish-shaped *chiwen* but instead the historical figures Zhang Fei with a spear and Yan Yan with a broadsword. Inside, three gilded plaques prominently display the inscriptions: "*Sansui Tang*" (Hall of Three Ears of Grain), "*Lingtai jingshi*" (The Founding of the Spirit Platform), and "*Chengshi shanlin*" (Oasis in the Heart of the City). These plaques highlight the hall's name, explain the purpose behind the garden's creation, and emphasize its design philosophy.

To the north of Sansui Hall lies the Grand Rockery, the centerpiece of Yu Garden. Standing approximately 14 meters tall, it is the largest surviving yellow stone rockery from the Ming dynasty in the Jiangnan region. Crafted by the renowned artisan Zhang Nanyang, who was celebrated for his expertise as both a painter and rockery designer, the rockery is admired for its majestic peaks and graceful valleys. Although climbing the rockery is no longer allowed, visitors can still appreciate its beauty from a distance, imagining the experience of standing amid its rugged landscape. At the summit of the rockery sits Wangjiang Pavilion, offering a panoramic view of the entire garden. While it may not seem particularly tall by today's standards, it was once a favored spot for locals to gaze out over the Huangpu River, especially popular during the Double Ninth Festival, which typically falls in October.

三穗堂与大假山
Sansui Hall and the Grand Rockery

上：三穗堂
U: Sansui Hall
下：大假山
D: The Grand Rockery

园景撷趣

Garden Scenery

听涛阁
Tingtao Belvedere

经过数次损毁和重建，豫园从最初的 70 余亩缩减至今天的近 30 亩，但因造景得当，既恢复了初建时"淡雅出之"的佳景，又有清乾隆（1736—1795）、嘉庆（1796—1820）时期所建的华丽建筑点缀其中，相得益彰。全园大小胜景四十余处，从格局上可分三穗堂、万花楼、点春堂、会景楼、玉华堂和内园六个景区，信步其中，可感布局之有致，设计之精妙。

Though Yu Garden has been reduced from its original over 70 *mu* to nearly 30 *mu* due to repeated destruction and reconstruction, its skillful landscape design preserves much of its original elegance while seamlessly incorporating splendid architecture from the Qianlong (1736–1795) and Jiaqing (1796–1820) periods. The garden boasts over 40 scenic spots, arranged into six distinct areas: Sansui Hall, Wanhua Tower, Dianchun Hall, Huijing Tower, Yuhua Hall, and the Inner Garden. A leisurely stroll through these areas reveals the garden's harmonious layout and intricate design, inviting visitors to appreciate its enduring beauty.

foreign merchants contributed to its upkeep, more guilds moved in, building walls and renovating structures, which disrupted parts of the gardens' original layout and landscape.

In 1842, during the First Opium War, British forces captured Shanghai and occupied West Garden, using it as a military camp, which caused significant damage. A decade later, during the Small Swords Society Uprising in 1853, the rebels used Yu Garden as their base, with Dianchun Hall serving as their northern command center. Other structures, such as Cuixiu Hall and Wanhua Tower, were repurposed as offices and storage. When the Qing army retook the city in 1855, they attacked West Garden, setting fire to Dianchun Hall, Xiangxue Hall, and other buildings, leaving the once-beautiful garden in ruins, with "the scenery washed away, and the springs and stones colorless." Between 1860 and 1862, as Taiping forces approached Shanghai, British and French troops jointly occupied West Garden as part of their military operations against the Taiping forces. The French soldiers drastically altered the landscape, filling ponds with stones and destroying much of the garden's original features. After the conflict, more guilds moved into Yu Garden, and by 1867, 21 guilds had established themselves there. Each guild restored different sections of the garden: the Bean and Rice Guild restored Sansui Hall, Cuixiu Hall, and Wanhua Tower; the Flower Sugar and Foreign Trade Guild restored Dianchun Hall; and the Cloth Guild restored Deyue Tower. These restoration efforts introduced a distinct Shanghai-style character to Yu Garden, which became known as the "Guild Garden" during this period.

During the 1937 Battle of Shanghai, West Garden sustained significant damage once again. Xiangxue Hall was destroyed by Japanese forces, and Yu Garden was designated as a refugee area, where makeshift shelters were erected, further degrading its scenery. Before 1949, most of the garden's structures were in severe disrepair. In 1956, a major restoration project began. As part of this effort, the Sacred Garden (also known as East Garden) within the City God Temple was linked to the West Garden, and together they became collectively known as Yu Garden. The area surrounding the Huxin Pavilion and Nine-bend Bridge was designated as a scenic area outside the main garden. In 1959, Yu Garden was designated a protected historical and cultural site in Shanghai and opened to the public in 1961. It was further recognized as a Major Historical and Cultural Site Protected at the National Level in 1982. Between 1986 and 1988, under the guidance of Professor Chen Congzhou, the eastern section of Yu Garden was reconstructed. Scenic areas like Yulinglong (Exquisite Jade Rock), Yuhua Hall, Huijing Tower, and Jiushi Gallery were restored, focusing on reshaping the landscape, dredging ponds, altering bridges, and planting new greenery to create a peaceful and elegant atmosphere. The eastern section took shape with winding waterways and pavilions artfully integrated into the landscape. An ancient opera stage, originally from the Finance Guild in former Zhabei District, was relocated to the southern side of the Inner Garden, enhancing Yu Garden's traditional leisure functions and adding spaces for state banquets and performances. In the early 21st century, new structures such as Hanbi Tower and Tingtao Belvedere were added, further refining the garden's functionality and contributing to the Yu Garden we know today.

古戏台
Ancient Opera Stage

in Shanghai pooled their resources to purchase significant portions of the garden, incorporating them into the City God Temple complex. Located to the northwest of the old City God Temple, the garden was renamed West Garden. Extensive renovations and expansions restored the garden to its original size of 70 *mu*, a process completed by 1784. This restoration included the construction of Sansui Hall, Cuixiu Hall, and the Huxin Pavilion, establishing much of the layout seen today. Entrusted to the Daoists of the City God Temple, West Garden became a hub for religious ceremonies, cultural events, and entertainment, transforming from a private retreat into a public space with religious, commercial, and cultural functions. The area around the Huxin Pavilion — connected to nearby buildings by a Zigzag Bridge — gradually developed into a bustling commercial hub, as storefronts were subdivided and leased, eventually becoming an integral part of the old city's marketplace. However, during the Jiaqing (1796–1820) and Daoguang (1821–1850) periods, the garden again fell into disrepair. As

九狮轩
Jiushi Gallery

进行东部重建工程，重修了玉玲珑、玉华堂、会景楼、九狮轩等景观，"叠山理水，疏池浚流，引廊改桥，栽花种竹，以空灵高洁之致为归"，令东部"园隔水曲，楼阁掩映，初具规模"[1]，又将原建于闸北钱业会馆的古戏台移建至内园南侧，让豫园既具备了传统的园林游憩功能，更增添了国宴迎宾、曲艺观赏的功能。21世纪初，豫园又增建了涵碧楼、听涛阁，使园林功能更为完善，这才造就了我们今日所见的豫园。

玉华堂内景
Interior of Yuhua Hall

Located at No.218 Anren Street in Shanghai's Huangpu District, Yu Garden spans nearly 30 *mu* (20 000 square meters) and stands as one of the city's most iconic gardens. It was commissioned by Pan Yunduan (1526–1601), a scholar who attained the highest rank (*jinshi*) in the imperial palace examination in 1562. Pan later served as the Right Provincial Administration Commissioner of Sichuan, managing provincial finances and communications. His family was distinguished for its academic success, with both his father and brother also attaining the *jinshi* titles, giving rise to the saying, "One family, three *jinshi*." In 1559, Pan began building the garden as a peaceful retreat for his father, who had retired from official duties. Pan Yunduan named it "Yu Garden," with "Yu" symbolizing peace and pleasure. While construction initially progressed slowly, it accelerated in 1577 when Pan retired and personally oversaw the project. By the late Wanli period, Yu Garden had expanded to over 70 *mu* (approximately 46 667 square meters) and was celebrated as "the most exquisite garden in Jiangnan" and "the crown of Jiangnan gardens."

In 1601, during the Wanli period of the Ming dynasty, after the death of Pan Yunduan, the Pan family's fortunes declined, and they struggled to maintain Yu Garden, leading to its gradual deterioration. During the Kangxi period (1662–1722) of the Qing dynasty, various guilds began using the garden's halls for meetings. In 1760, during the Qianlong period, local gentry

得月楼
Deyue Tower

园林面目全非。战争结束后,更多的同业公所进驻豫园,到1867年时已有21家。各公所自行修复了部分建筑,如豆米业公所修复三穗堂、萃秀堂、万花楼一区,花糖洋货业公所修复点春堂一区,布业公所修复得月楼一区,不同的修建风格和手法使豫园开始有了海派园林的风格特征。这一时期,豫园也被称为"公所园林"。

1937年淞沪会战中,西园又遭重创。香雪堂被日军焚毁,豫园被划为难民区,难民于其中搭建棚屋,园景再次被毁,至1949年前,园内建筑大多破败不堪。1956年,豫园进行大面积修缮,将城隍庙内的灵苑(又称"东园")和西园连在一起,总称为"豫园",并将湖心亭、九曲桥一带划为园外景观。1959年,豫园被公布为上海市文物保护单位,1961年向市民开放,1982年被公布为全国重点文物保护单位。1986—1988年,豫园在陈从周教授指导下

场所。清乾隆二十五年（1760），上海士绅集资购买豫园大片土地，归入邑城隍庙，因园位于老城隍庙之西北面，改称"西园"，并大加整治扩建，恢复至70余亩。此工程延续至乾隆四十九年（1784），三穗堂、萃秀堂以及湖心亭一带均为此时期建设，基本奠定了今日豫园之格局。由于托管给城隍庙道士，西园常举办宗教祭祀活动和文化娱乐活动，从最初的文人园林转变为兼具宗教、商业、文化、娱乐的城市公共空间，尤其是由曲桥连通周遭建筑的湖心亭一带，逐渐因店面分租、市肆兴盛而成为老城厢商业空间的组成部分。清嘉庆（1796—1820）、道光年间（1821—1850），园子再度颓废，因西园"为洋商捐施而成"（洋商指洋货商），更多的会馆公所借此入驻园内，在园中修筑围墙以划定界线，修缮房舍，破坏了部分格局和景观。

1842年，第一次鸦片战争时期，英军攻陷上海，占领西园作为兵营，对园林造成了较大的破坏。1853年

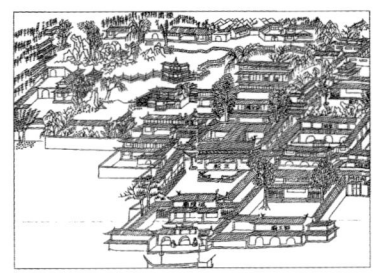

清同治《上海县志》城隍庙图
The City God Temple, *Gazetteer of Shanghai County, the Tongzhi Period* of the Qing dynasty

清末的湖心亭一带
The area around Huxin Pavilion in the late Qing dynasty

小刀会起义期间，起义军驻扎园中，将点春堂用作城北指挥部，萃秀堂、万花楼等用作办公、储物之地。1855年清军破城后攻打西园，点春堂、香雪堂等建筑付之一炬，一时"风光如洗，泉石无色"。1860—1862年，太平军逼近上海，英法联军联合抵御，西园成为法军屯兵之地，掘石填池，

历史沿革

Garden History

豫园位于黄浦区安仁街 218 号，现占地面积近 30 亩（20 000 平方米），是上海最具代表性的园林之一。豫园主人潘允端（1526—1601）是明嘉靖四十一年（1562）进士，官至四川右布政使，其父、其兄皆为进士，时有"一家父子三进士"之说。嘉靖三十八年（1559），潘允端为了让辞官告老还乡的父亲安享晚年，在位于安仁里（今安仁街，豫园东端）的祖宅西部修建园林，取名"豫园"，取"豫"字"平安""安泰"之意。豫园在建园的前十年进展缓慢，直到明万历五年（1577）潘允端解职归田、亲自督工才加快进度，于万历后期建成。当时的豫园占地 70 余亩（约 46 667 平方米），有"奇秀甲于江南""江南名园之冠"之誉。

万历二十九年（1601），潘允端去世，潘家日益衰败，难以承受巨大的维护开支，园林日渐倾颓。清康熙年间（1662—1722），逐渐有城内的同业公所占据园中厅堂作为集会

萃秀堂
Cuixiu Hall

大假山远景
Distant view of Grand Rockery

漫步在上海的市中心,我们可以在钢筋水泥之中探寻到一片江南风光。那便是豫园,一座有着 460 多年历史的江南园林。

Amid the bustling heart of Shanghai, nestled within the city's modern concrete and steel, lies a serene oasis: Yu Garden. For over 460 years, this historic retreat has preserved the timeless beauty of the Jiangnan landscape.

积玉水廊
Jiyu Waterside Corridor

海上名园，城市山林——豫园

A Historic Oasis in the Heart of the City: Yu Garden

年代：明代
地址：上海市黄浦区安仁街 218 号
占地面积：近 20 000 平方米
保护级别：全国重点文物保护单位

Era: Ming dynasty
Address: No.218, Anren Street, Huangpu District, Shanghai
Area: nearly 20 000 square meters
Protection level: National priority protected site

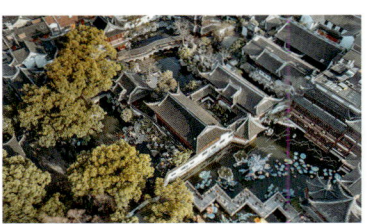

豫园鸟瞰图
Aerial View of Yu Garden

图例
Legends

1 三穗堂景区
  Sansui Hall Scenic Area

2 万花楼景区
  Wanhua Tower Scenic Area

3 点春堂景区
  Dianchun Hall Scenic Area

4 会景楼景区
  Huijing Tower Scenic Area

5 玉华堂景区
  Yuhua Hall Scenic Area

6 内园景区
  Inner Garden Scenic Area

7 上海城隍庙
  Shanghai City God Temple

8 湖心亭
  Huxin Pavilion

豫园

Yu Garden

1	**豫园**	Yu Garden
2	**秋霞圃**	Qiuxia Garden
3	**古猗园**	Guyi Garden
4	**醉白池**	Zuibaichi Garden
5	**曲水园**	Qushui Garden
6	**课植园**	Kezhi Garden

上海古园林分布示意图
Distribution Map of Shanghai Ancient Gardens

Yingxun Belvedere, Kezhi Garden

Features and Significance ······164	Features and Significance ······230
Historical Stories ······166	The Spirit of Studying and Tilling ······236

Relishing Scenic Beauty in Modern Times: Qushui Garden ······174
- Garden History ······176
- Garden Scenery ······183
- Yi Gate and Ninghe Hall ······184
- Xiyang Hongbanlou Tower and Deyue Gallery ······186
- Youjue Hall ······188
- Lotus Pond and Xiyu Bridge ······189
- Jiufeng Yilan Tower ······190
- Fogu Pavilion and Xiaohaoliang Pavilion ······191
- Tinglu Belvedere and Yaoyue Corridor ······192
- Jiyun Pavilion ······193
- Peony Pavilion and Lübo Corridor ······194
- Cultural Highlights ······196
- Flowers and Trees ······198

Cultivating Virtue and Legacy: Kezhi Garden ······204
- Garden History ······206
- Garden Scenery ······213
- Four Successive Halls ······214
- Stele Corridor ······218
- Yingxun Belvedere ······220
- Wangyue Tower ······221
- Daoshi Pavilion ······222
- Opera Stage ······224
- Shuijing Palace ······225
- Library Tower and Shuangbo Pavilion ······226
- Kezhi Bridge and Daoxiangcun ······228

Image Source ······239
Reference ······241
Epilogue ······244

Contents

Preface ·············13
Distribution Map of Shanghai Ancient Gardens
················23

A Historic Oasis in the Heart of the City:
Yu Garden ·············28
   Garden History ············30
   Garden Scenery ············39
   Sansui Hall and the Grand Rockery ·····40
   Wanhua Tower ············42
   Dianchun Hall and Hexu Hall ··········44
   Huijing Tower and Yulinglong ·········46
   Inner Garden ············48
   Features and Significance ·········50
   Gathering of Talents ············60

Over Five Centuries of Resilience: Qiuxia
Garden ·············68
   Garden History ············70
   Garden Scenery ············77
   Taohua Pool Scenic Area ·········78
   Ningxia Belvedere Scenic Area ·····84
   Qingjing Pond Scenic Area ········88
   City God Temple Scenic Area ·······91
   Autumn Scenery in Qiuxia Garden ·····94
   Cultural and Artistic Legacy ·······96

Flowing Waters and Timeless Charm: Guyi
Garden ·············100
   Garden History ············104
   Garden Scenery ············113
   Dhāraṇī Pillar of the Tang Dynasty ·····114
   South Hall ············116
   Putong Pagoda ············117
   Buque Pavilion ············118
   Weiyin Belvedere ············119
   Wan'an Pagoda ············120
   Cui'ai Tower ············121
   Luoyang Bridge ············122
   Xiaoyundou ············123
   Buxizhou ············124
   Xiaosonggang ············126
   Wulaofeng ············127
   Features and Significance ········128
   The Charm of Guyi Garden ········138

Poetic Elegance and Lasting Legacy: Zuibaichi
Garden ·············142
   Garden History ············144
   Garden Scenery ············151
   Outer Garden ············152
   Inner Garden ············154
   Collection of Notable Steles ········162

    碑刻艺术·····················162
    价值特色·····················164
    史海钩沉·····················166

怡景览胜，今朝传承——曲水园······174
    历史沿革·····················176
    园景撷趣·····················183
    仪门与凝和堂··················184
    夕阳红半楼与得月轩···············186
    有觉堂······················188
    荷花池与喜雨桥·················189
    九峰一览阁····················190
    佛谷亭与小濠梁·················191
    听橹阁与邀月廊·················192
    机云亭······················193
    牡丹亭与绿波廊·················194
    人文撷趣·····················196
    古木芳菲·····················198

课经植道，积厚流光——课植园······204
    历史沿革·····················206
    园景撷趣·····················213
    四进厅堂·····················214
    碑廊·······················218
    迎曛阁······················220
    望月楼······················221
    倒挂狮子亭····················222
    打唱台······················224
    水晶宫······················225
    藏书楼与双帛亭·················226
    课植桥与稻香村·················228

    价值特色·····················230
    课植精神·····················236

图片来源························239
参考文献························241
后记··························243

目录

前言 · 10

上海古园林分布示意图 · · · · · · · · · · · · · · · 23

海上名园,城市山林——豫园 · · · · · · 28
历史沿革 · 30
园景撷趣 · 39
三穗堂与大假山 · · · · · · · · · · · · · · · · · 40
万花楼 · 42
点春堂与和煦堂 · · · · · · · · · · · · · · · · · 44
会景楼与玉玲珑 · · · · · · · · · · · · · · · · · 46
内园 · 48
价值特色 · 50
人文荟萃 · 60

栉风沐雨,五百余载——秋霞圃 · · · · 68
历史沿革 · 70
园景撷趣 · 77
桃花潭景区 · 78
凝霞阁景区 · 84
清镜塘景区 · 88
城隍庙景区 · 91
秋霞"枫"景 · 94
艺文丰韵 · 96

曲水潆洄,海上秀园——古猗园 · · · · · · 100
历史沿革 · 104
园景撷趣 · 113
唐经幢 · 114
南厅 · 116
普同塔 · 117
补阙亭 · 118
微音阁 · 119
万安塔 · 120
翠霭楼 · 121
洛阳桥 · 122
小云兜 · 123
不系舟 · 124
小松冈 · 126
五老峰 · 127
价值特色 · 128
猗园风韵 · 138

古园诗韵,文脉流芳——醉白池 · · · · · · 142
历史沿革 · 144
园景撷趣 · 151
外园 · 152
内园 · 154

豫园两宜轩
Liangyi Gallery, Yu Garden

harmonious fusion of Chinese and Western styles.

However, they suffered extensive damage during Japan's invasion of China. In modern times, these gardens have been revitalized, blending with urban life and evolving alongside the city to regain their vibrancy. These ancient gardens are more than historical landmarks — they chronicle the city's evolution. Over the centuries, Qiuxia Garden, Guyi Garden, Yu Garden, and Zuibaichi Garden have endured dynastic changes, wartime devastation, and significant transformations. Initially created as classical residential gardens reflecting scholarly tastes, they later became temple gardens linked to the City God Temple and eventually transformed into local parks. Today, they serve as inclusive, open public spaces, exemplifying the evolution of Jiangnan gardens and showcasing Shanghai's unique approach to preserving this tradition. Shifts in purpose and audience led to variations in design philosophy and construction, resulting in some inconsistencies and imperfections. Nevertheless, the fusion of refined and popular elements in these traditional gardens makes them invaluable cultural assets, embodying Shanghai's historical legacy, spirit, and character.

As time passes, many gardens have faded into history — some preserved only in records, others immortalized in poetry and prose. Today, only a few ancient gardens remain in Shanghai, yet their timeless charm continues to captivate. This book features six exceptional gardens — Qiuxia Garden, Guyi Garden, Yu Garden, Zuibaichi Garden, Qushui Garden, and Kezhi Garden — selected for their artistic beauty, cultural significance, and accessibility. It aims to showcase their unique charm, celebrate their rich heritage, and advocate for the preservation of this invaluable legacy.

The timeless appeal of these ancient gardens embodies Shanghai's enduring spirit. Their openness, inclusivity, and innovative designs reflect the city's vibrant and dynamic character, making them integral to its cultural identity.

from Anhui began constructing Jie Garden. During the Wanli period (1573–1620), Zhu Zhizheng extensively remodeled it into Yi Garden, which later became one of the three renowned gardens associated with the Li family.

Meanwhile, the Jin family, spanning three generations, built Jin Family Garden north of the Gong Family Garden. These gardens emerged after the decline in garden-building during the early Ming dynasty — a consequence of the Hongwu Emperor's restrictive policies. Their development marked a revival in Shanghai's horticultural history.

In 1559, Pan Yunduan (1526–1601) of Shanghai County began constructing Yu Garden to the west of his family house in Anren Lane, creating a peaceful retreat for his retired father. In 1577, Pan resigned from his position as Right Provincial Administration Commissioner of Sichuan and returned home, personally supervising the accelerated construction of Yu Garden. Between the Wanli and Tianqi reigns, scholar Shen Hongzheng (1578–1627) built Shen Garden east of the Gong Family Garden. Together with Gong Family Garden and Jin Family Garden, it later formed today's Qiuxia Garden. At that time, Jiangnan was immersed in a flourishing wave of garden-building.

In the early Qing dynasty, Gu Dashen, Secretary in a Bureau of the Ministry of Works, retired in disillusionment to Yunjian (modern Songjiang District). On the former site of Guyang Garden, originally established by Zhu Zhichun in the Northern Song dynasty, he built a new garden centered around a pond, later naming it "Zuibaichi." In 1745, after careful site selection, the City God (Chenghuang) Temple in Qingpu began constructing Lingyuan Garden to its east, which was renamed Qushui Garden in 1798 during the Jiaqing reign.

By 1860, during the Taiping Heavenly Kingdom Movement, warfare reshaped the fate of Jiangnan gardens. Many in Shanghai suffered severe damage; although much of the original architecture was lost, the overall layouts largely survived. Following the urban development of the late Qing and Republic of China periods, Haipai gardens flourished, presenting a distinctive style marked by openness and inclusivity. In 1912, Ma Weiqi (1853–1928) of the Ma clan from Jingting, Zhujiajiao, began constructing Kezhi Garden — also known as the Ma Family Garden — at the northern end of Zhujiajiao ancient town. The garden blended classical traditions with modern influences, reflecting a

Preface

The gardens of Jiangnan seamlessly blend architecture, landscapes, and flora, offering a poetic connection between humanity and nature. Renowned for its abundant water resources and mild climate, the Jiangnan region has long been celebrated as the birthplace of classical Chinese garden art. These gardens showcase breathtaking natural beauty while reflecting the ingenuity of ancient scholars and craftsmen. Rooted in philosophy and inspired by traditional arts like poetry and painting, they exude artistic sophistication and hold a unique place in the history of global garden design.

Shanghai's ancient gardens, part of this rich tradition, embody the essence of Jiangnan's vibrant and multifaceted garden art. These gardens can be traced back to the period before the Tang dynasty, flourished during the Song and Yuan dynasties, peaked during the mid-to-late Ming dynasty garden-building boom, and continued to thrive into the early Qing dynasty. Although they began to decline in the mid-to-late Qing, they experienced a revival during the late Qing and Republic of China periods. During this time, some evolved into modern public parks and adopted the distinct Haipai garden style, reflecting Shanghai's "East Meets West" culture. Shanghai's unique modern history has imbued its gardens with a distinctive character, combining traditional craftsmanship with innovative design and blending openness with self-contained elegance. The evolution of these gardens mirrors Shanghai's social, economic, and cultural transformations, symbolizing the fortunes of families, the progress of society, and the unfolding chapters of history.

In 1478, during the Chenghua reign of the Ming dynasty, Gong Hong(1451–1526) attained the *jinshi* title by passing the imperial palace examination, launching his official career. By 1501, while serving as the Right Administration Vice Commissioner of Zhejiang, he retired due to illness and returned to his hometown in Jiading. During his mourning period and subsequent years of seclusion, Gong Hong began creating the Gong Family Garden. After formally retiring in 1521 with the honorary title of Minister of Works, he expanded the garden, completing it shortly before his death in 1526. Around the same time, in Nanxiang, over 10 kilometers from Jiading, the Min family

份转换,成为如今多元、开放的城市公共空间,这是江南园林嬗变的范式之一,也是上海传承江南园林且有别于其他地区的独特之处。在这一转换过程中,因服务对象与需求不同,设计理念与造园手法必然有所差异,也造成了一些矛盾与缺憾。然而,传统园林在雅俗纷呈、兼容并蓄的面貌中,构成了上海珍贵的城市历史文化资源,彰显着上海的精神与品格。

闲潭云影,物换星移,如今诸多园林湮没于历史长河中,或零落成史志上的只言片语,或传颂于文人的诗词歌赋之间。至今上海遗留的古园林不多,但仍异彩纷呈,本书选取了其中有代表性且对外开放、便于探访的六处,旨在展现园林艺术之美,弘扬其文化内涵,以便更好地保护和传承这一宝贵的文化遗产。

"名园依旧风流在,检点云山入画图",如今以秋霞圃、古猗园、豫园、醉白池、曲水园、课植园等为代表的上海古园林的魅力历久弥新,而海纳百川、兼容并蓄的城市精神已深入上海的"骨髓",体现着"开放、包容、创新"的上海城市精神品格。

了金氏园,这是上海摆脱明初"禁园"阴影后出现的江南园林。

明嘉靖三十八年(1559),上海县潘允端(1526—1601)为让告老还乡的父亲安享晚年,在位于安仁里的祖宅西部修建豫园。明万历五年(1577),潘允端从四川右布政使任上辞归,亲自督工,加快营造豫园。万历、天启年间,诸生沈弘正(1578—1627)在龚氏园东造沈氏园(后龚氏园、金氏园、沈氏园形成今日之秋霞圃)。彼时,整个江南沉浸在造园浪潮中,上海地区的营建之势蔚为壮观。

清初,工部主事顾大申宦途失意,归隐云间,于北宋朱之纯所建的"谷阳园"遗址之上营建池上云林,并以"醉白"命名。清乾隆十年(1745),经长期选址规划,青浦城隍庙于东侧始建灵园,清嘉庆三年(1798)更名为"曲水园"。

清咸丰十年(1860),太平军战事改变了江南诸园的存亡命运,上海园林未能幸免,此间多有损毁,幸而建筑虽损,园林格局犹存。清末民国时期,随着近代城市发展,海派园林开始蓬勃兴起,呈现出博采兼容的面貌。民国元年(1912),朱家角井亭马氏族人马维骐(1853—1928)开始在朱家角古镇北端建造"课植园",别称"马家花园",呈古今融合、中西合璧之风。

在近代日本侵华战争中,各园林多历兵火,受损严重。直至新时代,古园林融入现代生活,与城融合,与城俱进,焕发新生。古园林记录和承载了城市历史过程的变迁,是历史发展的特殊参照。回顾数百年间,秋霞圃、古猗园、豫园、醉白池等园林历经朝代更迭、战火破坏、变迁繁复,实现了从彰显文人趣味的古典宅园,至多元与世俗化的城隍庙园,再至地方公园的身

前言

中国园林汇聚建筑、山水、花木等多种艺术形式及审美情怀，是人类诗意栖居的象征。江南地区自古以来就以其丰富的水资源和温和的气候条件而著称，为古代园林艺术的发展提供了独特的环境。江南园林于沃土应运而生，巧妙地借鉴自然之美，融合古代文人和能工巧匠的智慧，蕴涵哲学思想及山水诗画等传统艺术，被赋予了艺术灵魂，展现了中华文化高超的艺术审美和独有的深邃意境，是中国古典园林的杰出代表，是世界园林艺术中的宝贵遗产。

江南园林分布广泛且特色鲜明，上海地区的古园林传承江南园林精华，呈现出多元、生动的特点，其萌芽于唐代之前，宋元时期蔚起，在明代中晚期的营造浪潮中逐渐兴盛，清初持续发展，清代中晚期渐趋衰落，至清末民国时期，随着近代公共花园的兴起，海派园林潮涌。上海地区特殊的近代发展历程，使上海古园林在实践中形成了传承发展、开放包容、自成一体的独特风格，其历程与江南园林的发展演变相互呼应，也与上海地区的社会、经济、文化变迁紧密相关。一座园林的兴衰，是一个家族兴衰变迁的体现，也是一个地区社会发展、历史演变的具象反映。

明成化十四年（1478），龚弘（1451—1526）荣登进士，始沉浮宦海。明弘治十四年（1501），龚弘自浙江右参政任上归乡养疾，之后丁忧在乡，居家十多年，在此期间完成了嘉定龚氏园主体部分的草创。明正德十六年（1521），龚弘以工部尚书衔致仕，至明嘉靖五年（1526）去世前，续造园林，此为龚氏园成形年代。同在嘉靖初年，二十多里外繁华的南翔，徽籍闵氏开始营建借园，后于明万历年间（1573—1620）经朱稚征擘画设计，借园成为李氏三园之一的"猗园"。数年间，豪族金氏祖孙三代在龚氏园北建造

| 建筑可阅读 |

Jiangnan from the Perspective of Cultural Heritage
Ancient Gardens in Shanghai

文物视角中的江南
上海古园林

上海市文物保护研究中心 编著
Shanghai Cultural Heritage Conservation and Research Center

同济大学出版社·上海
TONGJI UNIVERSITY PRESS·SHANGHAI

建筑可阅读书系

文物视角中的江南：
上海古园林

编著：上海市文物保护研究中心

编委会

主任：钟晓敏
副主任：向义海
编委：李晶、邓军、钟经纬、舒晟岚、翟杨

中文文字：高文虹、蒋薇、李弥、倪家
英文翻译：高毓婷
英文校对：李弥、蔡薛成、徐捷
摄影：许一凡，等

鸣谢（排名不分先后）：
我们衷心感谢以下单位对本书创作提供的宝贵支持。

上海豫园管理处
上海嘉定秋霞圃
上海古猗园
上海醉白池公园
上海曲水园
上海朱家角古镇旅游发展有限公司
上海市历史博物馆（上海革命历史博物馆）
上海市松江区文物保护管理所
上海市嘉定博物馆
上海市青浦区博物馆
上海市青浦区朱家角镇社区文化活动中心

Stories of Shanghai Architecture Series

Jiangnan from the Perspective of Cultural Heritage: Ancient Gardens in Shanghai

Shanghai Cultural Heritage Conservation and Research Center

Editorial Board

Director: Zhong Xiaomin
Deputy director: Xiang Yihai
Editorial members: Li Jing, Deng Jun, Zhong Jingwei, Shu Shenglan, Zhai Yang

Chinese text: Gao Wenhong, Jiang Wei, Li Mi, Ni Jia
English translation: Gao Yuting
English proofreading: Li Mi, Cai Xuecheng, Xu Jie
Photography: Xu Yifan, et al.

Acknowledgements (in no particular order):
We extend our heartfelt gratitude to the following organizations for their invaluable support in the creation of this book.

The Management of Shanghai Yu Garden
Shanghai Jiading Qiuxia Garden
Shanghai Guyi Garden
Shanghai Zuibaichi Garden
Shanghai Qushui Garden
Shanghai Zhujiajiao Ancient Town Tourism Development Co., Ltd
Shanghai History Museum (Shanghai Revolution Museum)
Shanghai Songjiang District Cultural Heritage Conservation Office
Shanghai Jiading Museum
Shanghai Qingpu District Museum
Zhujiajiao Community Cultural Activity Center, Qingpu District, Shanghai

古猗园不系舟
Buxizhou, Guyi Garden

Rockery of Qiuxia Garden

豫园大假山
Grand Rockery, Yu Garden